Materials Processing for Production of Nanostructured Thin Films

Materials Processing for Production of Nanostructured Thin Films

Editor

Keith J. Stine

MDPI • Basel • Beijing • Wuhan • Barcelona • Belgrade • Manchester • Tokyo • Cluj • Tianjin

Editor
Keith J. Stine
University of Missouri-Saint Louis
USA

Editorial Office
MDPI
St. Alban-Anlage 66
4052 Basel, Switzerland

This is a reprint of articles from the Special Issue published online in the open access journal *Processes* (ISSN 2227-9717) (available at: https://www.mdpi.com/journal/processes/special_issues/Nanostructured_Thin_Films).

For citation purposes, cite each article independently as indicated on the article page online and as indicated below:

LastName, A.A.; LastName, B.B.; LastName, C.C. Article Title. *Journal Name* **Year**, *Volume Number*, Page Range.

ISBN 978-3-0365-1088-0 (Hbk)
ISBN 978-3-0365-1089-7 (PDF)

© 2021 by the authors. Articles in this book are Open Access and distributed under the Creative Commons Attribution (CC BY) license, which allows users to download, copy and build upon published articles, as long as the author and publisher are properly credited, which ensures maximum dissemination and a wider impact of our publications.

The book as a whole is distributed by MDPI under the terms and conditions of the Creative Commons license CC BY-NC-ND.

Contents

About the Editor vii

Keith J. Stine
Special Issue "Materials Processing for Production of Nanostructured Thin Films"
Reprinted from: *Processes* 2021, 9, 298, doi:10.3390/pr9020298 1

Malik Muhammad Nauman, Murtuza Mehdi, Dawood Husain, Juliana Haji Zaini, Muhammad Saifullah Abu Bakar, Hasan Askari, Babar Ali Baig, Ahmed Ur Rehman, Hassan Abbas, Zahid Hussain and Danial Zaki
Stretchable and Flexible Thin Films Based on Expanded Graphite Particles
Reprinted from: *Processes* 2020, 8, 961, doi:10.3390/pr8080961 5

Xiaofei Sheng, Yajuan Cheng, Yingming Yao and Zhe Zhao
Optimization of Synthesizing Upright ZnO Rod Arrays with Large Diameters through Response Surface Methodology
Reprinted from: *Processes* 2020, 8, 655, doi:10.3390/pr8060655 15

Ohud S. A. ALQarni, Riadh Marzouki, Youssef Ben Smida, Majed M. Alghamdi, Maxim Avdeev, Radhouane Belhadj Tahar and Mohamed Faouzi Zid
Synthesis, Electrical Properties and Na+ Migration Pathways of $Na_2CuP_{1.5}As_{0.5}O_7$
Reprinted from: *Processes* 2020, 8, 305, doi:10.3390/pr8030305 29

Petr Filip and Petra Peer
Characterization of Poly(Ethylene Oxide) Nanofibers—Mutual Relations between Mean Diameter of Electrospun Nanofibers and Solution Characteristics [†]
Reprinted from: *Processes* 2019, 7, 948, doi:10.3390/pr7120948 47

Yujing Zhang and Chih-Hung Chang
Metal–Organic Framework Thin Films: Fabrication, Modification, and Patterning
Reprinted from: *Processes* 2020, 8, 377, doi:10.3390/pr8030377 57

Jay K. Bhattarai, Md Helal Uddin Maruf and Keith J. Stine
Plasmonic-Active Nanostructured Thin Films
Reprinted from: *Processes* 2020, 8, 115, doi:10.3390/pr8010115 103

About the Editor

Keith J. Stine, Professor of Chemistry, Department of Chemistry, University of Missouri—Saint Louis, MO 63121; Keith J. Stine received his BS from Fairleigh Dickinson University in 1984 and his Ph.D. in 1988 from MIT under the direction of Carl W. Garland. He was a postdoctoral fellow at UCLA in the lab of Charles M. Knobler. He joined the faculty at the University of Missouri—Saint Louis in 1990. Dr. Stine's research effort involves studies of modified surfaces and nanostructures. The surface modification of nanostructures is pursued with a focus on their prospective applications in bioanalytical chemistry, such as in immunoassays, sensors, or in separations. Other projects concern the study of lipid monolayers and bilayers as models of processes occurring at the surface of cell membranes, and the use of these monolayers in molecular recognition studies.

Editorial

Special Issue "Materials Processing for Production of Nanostructured Thin Films"

Keith J. Stine

Department of Chemistry and Biochemistry, University of Missouri-Saint Louis, Saint Louis, MO 63121, USA; kstine@umsl.edu; Tel.: +1-314-516-5346

Citation: Stine, K.J. Special Issue "Materials Processing for Production of Nanostructured Thin Films". *Processes* **2021**, *9*, 298. https://doi.org/10.3390/pr9020298

Received: 28 January 2021
Accepted: 29 January 2021
Published: 4 February 2021

Publisher's Note: MDPI stays neutral with regard to jurisdictional claims in published maps and institutional affiliations.

Copyright: © 2021 by the author. Licensee MDPI, Basel, Switzerland. This article is an open access article distributed under the terms and conditions of the Creative Commons Attribution (CC BY) license (https://creativecommons.org/licenses/by/4.0/).

The field of thin film technology [1] dates back many decades and has led to applications in areas such as display technology, the development of surfaces with desirable optical reflectance properties, coatings of medical devices for biocompatibility, corrosion protection, semiconductor device fabrication, polymer coatings for tuning wettability, coatings for providing hardness or protection, piezoelectric transducers, photovoltaic films for solar panels, chemical and biological sensors, and other areas. Many of the techniques for production of thin films, typically in the microns range in thickness, were developed in earlier decades such as electrodeposition, electroplating, spin-coating, dip coating, vacuum sputtering, chemical vapor deposition, physical vapor deposition, thermal evaporation, atomic layer deposition, molecular beam epitaxy, electron beam deposition, pulsed laser deposition and other advanced technologies that have driven the modern age. When combined with advances in lithography into the extreme ultraviolet [2], and newer approaches such as microcontact printing [3], thin film technology has made possible many of the advanced devices introduced in the past decade. New methods such as molecular imprinting [4] are open for further investigation.

The emergence and rapid growth of nanomaterials science throughout the period since the 1980s, has added an entirely new dimension to thin film technology in the realization that introducing nanostructured features into thin films in an intentional and designed manner can impart novel and useful properties and also alter existing properties for films prepared by traditional methods. The Special Issue on "Materials Processing for Production of Nanostructured Thin Films" presents a number of articles and reviews in which the properties and applications provided by the introduction of features on the nanometers scale into thin films are described and investigated. The introduction of nanoscale features represents a new frontier in thin film technology that will have major impacts in the coming decades in industries ranging from catalysis, renewable energy, sensor technology, device technology, biotechnology and others.

The paper by Nauman et al. [5] demonstrates that nanostructures can be embedded into flexible substrates for use in the growing field of flexible electronics. In this study, bath sonication is used to cause the intercalation of expanded graphite particles into polydimethylsiloxane (PDMS) substrates formed in a mold. Sonication was used to intercalate silver or silver oxide into the graphite particles, while electrolysis was used to intercalate sulfate anions. The graphite modified PDMS could flex 1000 times and remain functional and with favorable mechanical properties.

The paper by Sheng et al. [6] helps to illustrate how the introduction of nanostructures and the geometric features of the nanostructures in a thin film can depend on the adjustment of multiple parameters in the production process. In this paper, vertically oriented zinc oxide nanorods are formed on sapphire substrates with large diameters using chemical bath deposition. Such arrays of ZnO on substrates have applications in solar cells and other technologies. The variables of precursor concentration, reaction temperature, reaction time, and ratio of Zn^{2+} to citrate anion influence the diameter of the nanorods in a way that was able to be modeled using the statistical approach of response surface methodology. The approach was more efficient than changing one variable at a time and used experiments in which multiple variable were changed. Spin-coating was also involved in priming the sapphire substrates with a thin layer of zinc oxide.

The paper by AlQarni et al. [7] describes the discovery of a new alkali metal phosphate material with the novel property of fast ionic conduction. The new material, $Na_2CaP_{1.5}As_{0.5}O_7$ was produced by the heating and grinding of precursors to produce a polycrystalline powder. The characterization methods of X-ray powder diffraction (XRD), Fourier transform infrared spectroscopy (FTIR), differential scanning calorimetry (DSC), energy-dispersive X-ray spectroscopy (EDX), scanning electron microscopy (SEM), and impedance spectroscopy were used to provide insight into how the material structure gave rise to fast ionic conduction as explained using a bond-valence site energy (BVSE) model.

The paper by Filip et al. [8] also illustrates how the production of a nanostructured thin film can depend on a complex set of variables in the case of forming electrospun layers of polyethylene (PEO). These electrospun fibers that form a nonwoven textile can have applications as filters, in tissue engineering, drug delivery, and other fields. In this study, the correlations between the variables of polymer molecular weight, polymer concentration and solution viscosity with nanofiber diameter are established for electrospinning from solutions in distilled water.

The review by Zhang et al. [9] provides a comprehensive survey of the formation of thin films of metal–organic frameworks (MOF) that are versatile structures formed by the coordination of multidentate organic linkers to metal cations. MOFs provide access to tenability of pore size and have found use in gas storage, catalysis and sensor development. A wide range of thin film formation methods are covered in this review including spin-coating, dip-coating, template synthesis, layer by layer deposition, evaporation-induced crystallization, hydro/solvothermal synthesis, electrochemical synthesis, atomic layer deposition, chemical vapor deposition and physical vapor deposition, and some others. The variety of possible substrates on which MOF thin films can be formed ranges from Si, SiO_2, to metals, metal oxides and nylon. The review emphasizes the role of surface pretreatment with self-assembly monolayers, and the possibilities for postsynthesis modification and patterning of the MOF thin films. In addition to the tuning of pores by the selection of metal cations and linkers, the preparation conditions determine the nanostructured texture or pattern of the MOF thin films.

The review by Bhattarai et al. [10] describes the growing field of nanostructured thin films with plasmonic properties such as localized surface plasmon resonance, propogating surface plasmon resonance and surface-enhanced Raman responses. The tuning of the nanoscale features of these films, referred to as plasmonic active thin films (PANTFs), can vary the plasmon wavelengths and refractive index sensitivity when used in biosensor applications. Methods for their production including electron beam lithography, nanosphere lithography, focused ion beam milling, porous membrane lithography and others used to make a wide variety of PANTFs are described. The creation of PANTFs with nanoscale gaps and nanoholes is described, along with those presenting arrays or patterns of elevated features such as nanodomes or nanopillars.

It is hoped that the papers in the Special Issue will inspire further development of nanostructured thin films by new and creative approaches. The Guest Editor thanks Ms. Tami Hu for her dedicated and generous assistance in the development of this Special Issue.

Funding: This research received no external funding.

Conflicts of Interest: The author declares no conflict of interest.

References

1. Frey, H.; Khan, H.R. (Eds.) *Handbook of Thin-Film Technology*; Springer: Berlin/Heidelberg, Germany, 2015. [CrossRef]
2. Fu, N.; Liu, Y.; Ma, X.; Chen, Z. EUV Lithography: State-of-the-Art Review. *J. Microelectron. Manuf.* **2019**, *2*, 19020202. [CrossRef]

3. Bhave, G.; Gopal, A.; Hoshino, K.; Zhang, J.X. Microcontact Printing. In *Encyclopedia of Nanotechnology*; Bhushan, B., Ed.; Springer: Dordrecht, The Netherlands, 2015. [CrossRef]
4. Chen, L.; Wang, X.; Lu, W.; Wu, X.; Li, J. Molecular imprinting: Perspectives and applications. *Chem. Soc. Rev.* **2016**, *45*, 2137–2211. [CrossRef] [PubMed]
5. Nauman, M.M.; Mehdi, M.; Husain, D.; Zaini, J.H.; Abu Bakar, M.S.; Askari, H.; Ali Baig, B.; Ur Rehman, A.; Abbas, H.; Hussain, Z.; et al. Stretchable and Flexible Thin Films Based on Expanded Graphite Particles. *Processes* **2020**, *8*, 961. [CrossRef]
6. Sheng, X.; Cheng, Y.; Yao, Y.; Zhao, Z. Optimization of Synthesizing Upright ZnO Rod Arrays with Large Diameters through Response Surface Methodology. *Processes* **2020**, *8*, 655. [CrossRef]
7. ALQarni, O.S.A.; Marzouki, R.; Ben Smida, Y.; Alghamdi, M.M.; Tahar, R.B.; Zid, M.F. Synthesis, Electrical Properties and Na$^+$ Migration Pathways of Na$_2$CuP$_{1.5}$As$_{0.5}$O$_7$. *Processes* **2020**, *8*, 305. [CrossRef]
8. Filip, P.; Peer, P. Characterization of Poly(Ethylene Oxide) Nanofibers—Mutual Relations between Mean Diameter of Electrospun Nanofibers and Solution Characteristics. *Processes* **2019**, *7*, 948. [CrossRef]
9. Zhang, Y.; Chang, C.-H. Metal–Organic Framework Thin Films: Fabrication, Modification, and Patterning. *Processes* **2020**, *8*, 377. [CrossRef]
10. Bhattarai, J.; Maruf, M.H.U.; Stine, K.J. Plasmonic-Active Nanostructured Thin Films. *Processes* **2020**, *8*, 115. [CrossRef]

Article

Stretchable and Flexible Thin Films Based on Expanded Graphite Particles

Malik Muhammad Nauman [1],*, Murtuza Mehdi [2], Dawood Husain [3], Juliana Haji Zaini [1], Muhammad Saifullah Abu Bakar [1], Hasan Askari [4], Babar Ali Baig [5], Ahmed Ur Rehman [2], Hassan Abbas [2], Zahid Hussain [2] and Danial Zaki [2]

[1] Faculty of Integrated Technologies, Universiti Brunei Darussalam,
 Bander Seri Begawan BE 1410, Brunei Darussalam; juliana.zaini@ubd.edu.bn (J.H.Z.);
 saifullah.bakar@ubd.edu.bn (M.S.A.B.)
[2] Mechanical Engineering Department, NED University of Engineering & Technology,
 Karachi 75270, Pakistan; drmurtuza@neduet.edu.pk (M.M.); ahmed.khan527@outlook.com (A.U.R.);
 hassanabbas.12@hotmail.com (H.A.); zahid_ned@ymail.com (Z.H.); mdanialzaki1995@hotmail.com (D.Z.)
[3] Textile Engineering Department, NED University of Engineering & Technology, Karachi 75270, Pakistan;
 dawood@neduet.edu.pk
[4] School of Mechanical Engineering, Chung-Ang University, Dongjak-Gu Seoul 156756, Korea;
 hasanaskari@cau.ac.kr
[5] Mechanical Engineering Department, Pakistan Institute of Engineering & Applied Sciences,
 Islamabad 45650, Pakistan; babar2411@gmail.com
* Correspondence: malik.nauman@ubd.edu.bn

Received: 8 July 2020; Accepted: 3 August 2020; Published: 10 August 2020

Abstract: Stretchable and flexible graphite films can be effectively applied as functional layers in the progressively increasing field of stretchable and flexible electronics. In this paper, we focus on the feasibility of making stretchable and flexible films based on expanded graphite particles on a polymeric substrate material, polydimethylsiloxane (PDMS). The expanded graphite particles used in this work are prepared by utilizing bath sonication processes at the ultrasonic frequency of either the commercially available graphite flakes or graphite particles obtained through electrolysis under the interstitial substitution of silver and sulfate, respectively. The X-ray diffraction (XRD) patterns confirm that, due to the action of the bath sonication intercalation of graphite taking place, the resistances of the as-fabricated thin films is ultimately lowered. Mechanical characterizations, such as stretchability, flexibility and reliability tests were performed using home-made tools. The films were found to remain stretchable up to 40% tensile strain and 20% bending strain. These films were also found to remain functional when repeatedly flexed up to 1000 times.

Keywords: expanded graphite; flexible; polydimethylsiloxane; stretchable; thin films

1. Introduction

Graphite has found many technologically important applications. For instance, graphite particles have been used in thin film batteries, graphene synthesis, gas sensing, filter material, energy storage, steam generation and thin film nano composites [1–8]. Additionally, a progressively growing area where graphite thin films could be exploited is stretchable and flexible smart electronics. The most demanding features of smart electronics lies in the fact that thin layers of different materials built on such platforms must retain their functionalities when the device is mechanically stretched and/or flexed [9–12]. Hence, it has become exceedingly important to study the applicability of carbon materials and their performance within a mechanically stretchable and flexible framework.

Graphite is a layered allotrope of carbon in which each layer is composed of carbon atoms that are covalently attached to each other in an in-plane hexagonal fashion, whereas the layers are also weakly connected through van der Waals bonds along the lateral direction [13]. Here, it is worthy to note that most smart electronic devices will require at least one or two electrically conductive material layers for their functionality [14,15]. Therefore, in the context of nanostructured thin films, it will be highly desirable to somehow enhance the electrical conductivity of graphite particles used to fabricate the films. Theoretically speaking, if a single layer of graphite can be separated from its bulk counterpart, converting graphite into a two-dimensional planar material (graphene) its electrical conduction can be remarkably increased [16]. This process is known as the exfoliation of graphite. However, the removal of every single layer of graphite is not possible and can be heavily time consuming and expensive. Another method that is cost effective and much simpler to apply is the intercalation of graphite particles [17–19]. Intercalation refers to a technique of causing an increase in the inter-layer distance of graphite by inserting interstitial atoms or ions between the layers. Therefore, keeping in view the time and cost, two routes have been selected for this work. The first route consists of simply mixing the interstitial atoms with the commercially available graphite flakes and generating the intercalation effect using the bath sonication process. The second route is based on the sonication of graphite particles that are collected through the electrolysis of bulk graphite in aqueous electrolyte containing the interstitial atoms [20]. Previous work on graphite thin films was indeed important as far as electrical, thermal and/or morphological properties are concerned [21–23]. However, these vacuum-based fabrication methods require high setup cost and are also energy demanding. For this reason, it is anticipated that thin graphite films could be fabricated in a cost-effective manner and properties such as stretchability and flexibility, which remain of outmost importance for the realization of smart electronics, should also be addressed.

Therefore, the objective of this paper is to utilize the as-synthesized expanded graphite particles in the fabrication of thin graphite films and secondly to test the mechanical stretchability and flexibility of the as-fabricated graphite films on an intrinsically flexible and stretchable polymer substrate.

2. Materials and Methods

2.1. Materials and Chemicals

Commercial graphite flakes of average size 4.13 µm were purchased from Luoyang Tongrun Technology China. The graphite electrode and all chemicals, such as silver nitrate salt ($AgNO_3$), N,N-dimethylformamide also known as DMF, ammonium sulfate salt (($NH_4)_2SO_4$), ethylene glycol, acetone, deionized water and distilled water, were locally purchased from Fanara Chemicals and Supplies, Pakistan. PDMS Sylgard 184 elastomer kit was purchased from Dow Corning, Michigan, USA. All chemicals and materials were used as received without any further treatment.

2.2. Synthesis of Graphite Particles Using Process of Bath Sonication (First Route)

In this process, silver nitrate and commercial graphite flakes in a 1:1 ratio by weight were thoroughly stirred in deionized water using a magnetic stirrer until a black colored suspension was formed. The glass bottle containing the suspension was then placed in a bath sonicator containing distilled water and sonication was continuously applied for 4 h at a frequency of greater than 20 kHz. After sonication, the suspension was filtered and washed using deionized water. Finally, the residue collected over the filter paper was dried on a hot plate at 100 °C for 20 min to render the so-called expanded graphite particles. Figure 1 depicts the schematic of this process.

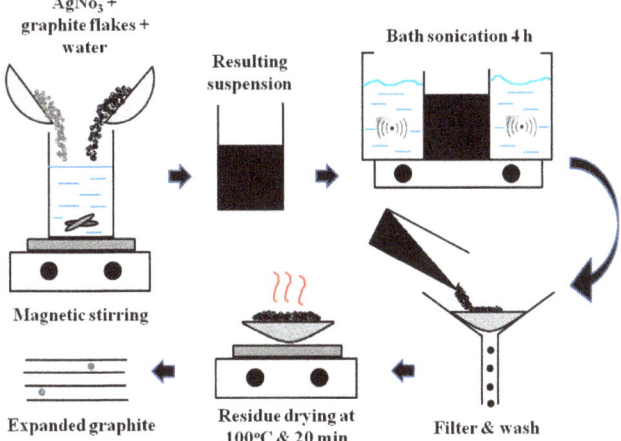

Figure 1. Schematic of first route. Arrows indicate the step by step procedure for obtaining expanded graphite particles using bath sonication.

2.3. Synthesis of Graphite Particles Using Process of Electrolysis (Second Route)

In this process a two-electrode electrochemical cell was used for the synthesis of expanded graphite particles. The graphite plate was made the anode while the copper plate was used as a cathode. The electrolyte was 500 mL 1 M solution of ammonium sulfate salt ((NH_4)$_2SO_4$) in distilled water. The process was carried out for 2 h at a voltage of 10 V with a constant current of 2 A. During this process, the graphite particles exfoliated away from the anode and kept on dispersing within the electrolyte. The electrolyte containing the suspended graphite particles was bath sonicated for 2 h followed by filtering and washing using deionized water. Finally, the residue was collected and dried on a hot plate at 100 °C and 20 min to render the expanded graphite particles. Figure 2 shows the schematic diagram of this process.

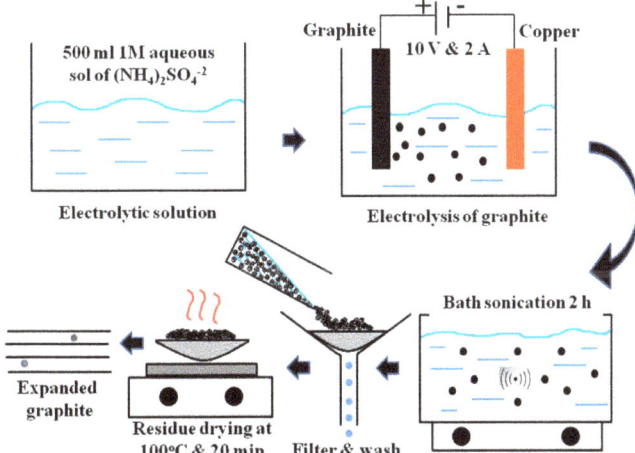

Figure 2. Schematic of second route. Arrows indicate the step by step procedure for obtaining expanded graphite particles using electrolysis of graphite.

2.4. Synthesis of Ink, PDMS Substrate and Expanded Graphite Thin Films

The ink containing the collected graphite particles for both the cases discussed above was prepared by thoroughly mixing 1.5 gm of the collected particles in 10 mL of DMF. DMF acts as a solvent for the ink and also as a conductive binder between expanded graphite particles within the thin films. In this work, the as-synthesized ink was thoroughly shaken each time before the fabrication of graphite films. In order to prepare PDMS substrates, the base and the cross linker were mixed together thoroughly in a ratio of 20:1 by weight. The solution was centrifuged at high rpm until all the air bubbles were removed. Molds were prepared using glass slabs that were 5 cm squared in size. Scotch tape was attached on one surface of the glass slabs for easy removal of the PDMS and the sides were secured using pieces of cardboard and scotch tape. Then, 2.5 mL of clear PDMS solution was poured at the center of the molds and left to cure at room temperature for 2 to 3 days, after which the PDMS substrates were cut in dimensions of 1 cm (width) × 4 cm (length) × 2 mm (thick). The as-fabricated PDMS substrates were found to remain reasonably stretchable and flexible for testing the films. In this work thin films of graphite on stretchable and flexible PDMS substrates were simply prepared by using a paint brush technique until the substrates were uniformly coated with the ink. The films were left to cure at room conditions for 24 h before performing the mechanical tests. By studying various lateral scanning electron micrographs (SEM) of the as-fabricated laminate, the average thickness of the graphite films in this work was found to remain 15 ± 1.37 µm. Figure 3 depicts a typical image of the film thickness.

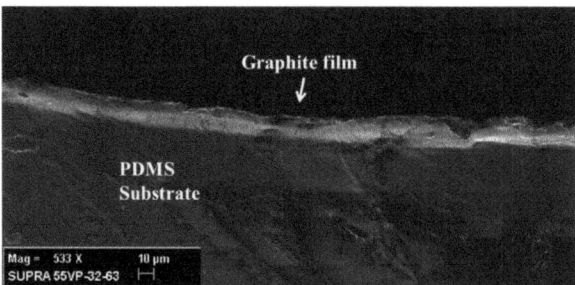

Figure 3. Lateral scanning electron microscope (SEM) image of a typical graphite thin film. The lateral cross-sectional image was viewed at an angle of 90 degrees.

2.5. Characterizations

From a material characterization point of view X-ray diffraction was performed for the collected expanded graphite particles in order to ensure the graphite phase and possible traces of other materials, such as silver and sulfate ions. The film resistances were measured using a digital multimeter with an accuracy of ±1%. Three types of mechanical characterizations were performed on the as-fabricated films. Firstly, the mechanical stretchability was characterized using a homemade thin film tensile tester. In this test the graphite/PDMS laminate was secured between the fixed and moveable clamps of the apparatus and it was stretched slowly by giving deformation in steps of 0.5 mm. After each step the resistance of the film was recorded. The stretchability testing was continued until the multimeter registered an open circuit. The second characterization was performed to quantify the mechanical flexibility of the as-fabricated films. In this test, the graphite/PDMS laminates were conformally wrapped around circular objects of varying diameters and the film resistance was recorded. It is to be noted that the tensile test induces tensile strain in the films, whereas the flexibility test induces bending strain in the films. The third mechanical characterization used in this study was rather to check the reliability of the as-prepared films. In this test a typical graphite/PDMS laminate was secured between the thumb and the index finger of the experimentalist and was pressed and then quickly depressed multiple times until the film was found to fail. The film resistance was recorded intermittently during the reliability tests. Figure 4 depicts all types of mechanical characterizations used in this work.

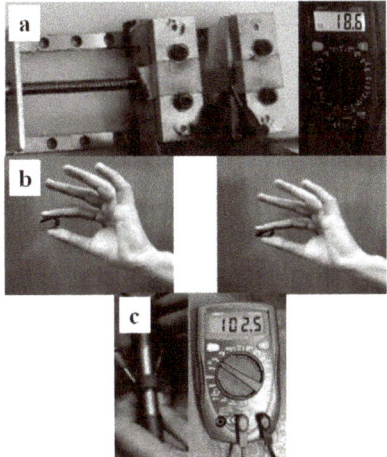

Figure 4. Various mechanical tests performed on the as-fabricated graphite/PDMS laminate. (**a**) Tensile test, (**b**) repeated reliability testing, and (**c**) bending test.

3. Results and Discussion

3.1. Material and Morphology of Expanded Graphite Particles

Figure 5 represents the XRD pattern of the expanded graphite flakes obtained by following the bath sonication process of commercial graphite flakes—i.e., the first route. In Figure 5 it can be seen that, as compared to the pure graphite, which is known to exhibit a characteristic peak at 26.53°, the XRD pattern of the expanded graphite flakes have also shown peaks at 32° and 38°. After comparing with various peaks of silver and its compounds, it was concluded that these new peaks refer to silver oxide and silver, respectively. This confirms the presence of silver and its compound in the expanded graphite flakes during the bath sonication process. SEM images shown in Figure 5 as insets also confirm the fact that, after performing the bath sonication process, the graphite flakes have become swollen and increased in overall size, which indicates that the graphite layers have, indeed, expanded during the process. This mechanism was ultimately found to reduce the film resistances by three orders of magnitude, as compared to the films of pristine graphite particles.

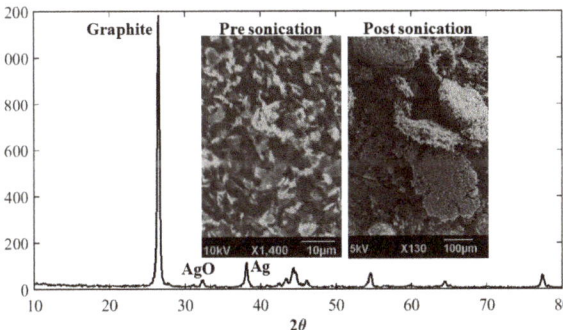

Figure 5. X-ray diffraction pattern of expanded graphite particles prepared by first route. The insets show the scanning electron micrograph (SEM) images of the expanded graphite particles before and after bath sonication.

On the other hand, Figure 6 shows the XRD pattern of expanded graphite particles collected through the process of electrolysis—i.e., the second route. Compared to the XRD pattern of pure graphite, several other characteristic peaks can be observed. These new peaks are due to the presence of sulfate ions. The peaks appear at angles of 16.87°, 20.43° and 22.78° with higher inter-atomic spacing of 5.25 Å, 4.34 Å and 3.90 Å, respectively. Higher inter-atomic spacing in the sample suggests that intercalation by the combined action of bath sonication and sulfate ions during electrolysis have been achieved. Again, it can be seen from the SEM images given in Figure 6 as insets that, after performing the bath sonication, the particle size has indeed increased, which qualitatively indicates the intercalation effect. The film resistances were lowered by three orders of magnitude, as compared to the films of pristine graphite particles. However, in this case it was noted that the film resistances on average remained five times lower than the film resistances prepared by the first route. This decrease in the film resistance is attributed to higher d-spacing caused by the sulfate ion as compared to the d-spacing caused by the intercalation of silver, which remains at 2.77 Å and 2.37 Å at angles of 32° and 38°, respectively.

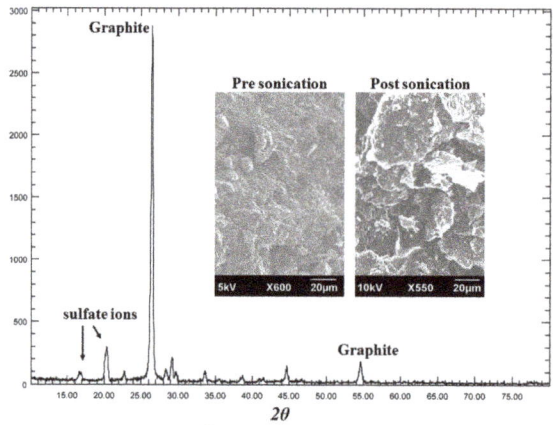

Figure 6. X-ray diffraction pattern of expanded graphite particles prepared by second route. The insets show the scanning electron micrograph (SEM) images of the expanded graphite particles before and after bath sonication.

3.2. Stretchability, Flexibility and Reliability of Expanded Graphite Thin Films

Figure 7 depicts the change in normalized resistance of the films prepared by following the first route and the second route against the applied tensile strains. It can be seen that expanded graphite films prepared by following the second route are approximately 1.3 times more stretchable than the first route. This is due to the fact that expanded graphite particles synthesized by following the second route were exfoliated from a hard graphite plate as an anode. Owing to this reason, the films formed by these expanded graphite particles were slightly higher in strength and therefore could bare larger tensile strains before failure. On the other hand, it could be observed that, up to 25% tensile strains, both films showed almost linear behavior and the increase in resistance of the films remains less than 10 in both cases. Specially, the films prepared by the first route showed even less increases in resistance. Thin film piezoresistive coefficient can be considered as a parameter that highlights the dependency of the resistance on resistivity changes due to mechanical loads. In order to estimate the gage factor and thus the piezoresistive coefficient for the as-fabricated films the experimental data shown in Figure 7 could be used in combination with the equation of the form [$\frac{\Delta R}{R_o} = \frac{R}{R_o} - 1 = (1 + \vartheta + \beta)\varepsilon = GF.\varepsilon$] [24], where ϑ is the Poisson ratio of graphite films taken to be 0.2 [25], R is the final film resistance, R_o is the initial film resistance, β is the piezoresistive coefficient of the film material, GF is the gage factor and ε is the applied tensile strain. Using the values from Figure 7 it can be estimated that the gage factors

GF for the as-fabricated films comes out to be 9.9 and 37.33 for the first route and the second route, respectively. Therefore, the piezoresistive coefficients are 8.7 and 36.13 for the first route and second route, respectively. This suggests that the change in resistivity of the material during mechanical loading affects the films prepared by the second route more strongly, which is the reason for higher resistance change during stretching for these films. These results fairly suggest the applicability of expanded graphite thin films in stretchable electronic devices.

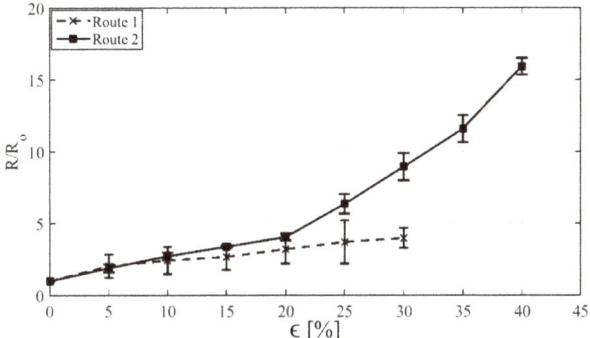

Figure 7. Stretchability curves of expanded graphite thin films. Error bars show standard deviations between the tested samples.

Figure 8 shows the variation in normalized resistance against the bending diameters. This graph can be considered as the quantitative measure of the flexibility of as-fabricated expanded graphite thin films. It is to be noted that, in this scenario, unlike the case of stretchability, the as-fabricated films are subjected to increasing bending strains. From Figure 8 it can be noticed that both the films remain functional until they are conformally wrapped around a circular rod of 10 mm diameter, which translates to a maximum bending strain of 20% using the equation of the form [$\varepsilon = \frac{t_f + t_s}{D}$] where t_f is the average film thickness, t_s is the thickness of the substrate and D is the bending diameter [26]. High gage factors of 49.3 and 89.2 for the films prepared by the first route and the second route, respectively, are estimated. This suggests that the as-fabricated films are capable of bending and can be flexibly operated up to a maximum of 20% bending strains with high strain sensitivity.

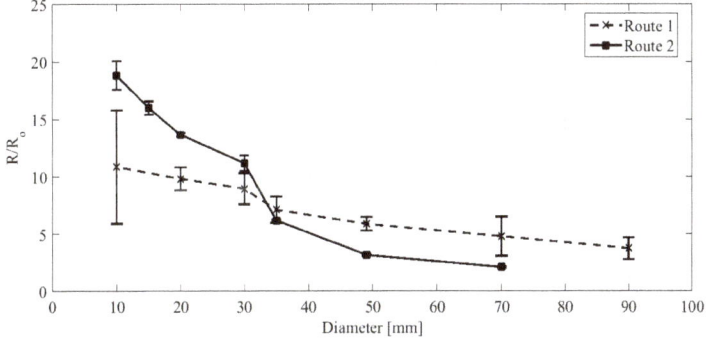

Figure 8. Bending curves indicating the flexibility of expanded graphite thin films. Error bars show standard deviations between the tested samples.

In order to conclude, the mechanical characterizations reliability tests were performed on the as-fabricated films and the results were shared in Figure 9. This figure shows the change in normalized resistance of the as-fabricated films against the number of flexure cycles. Both the films were found to withstand repeated flexure cycles, with films prepared by the second route slightly over performing the films prepared by the first route. However, the film resistance changed by 41.57 and reached a value of 46.97 kΩ.

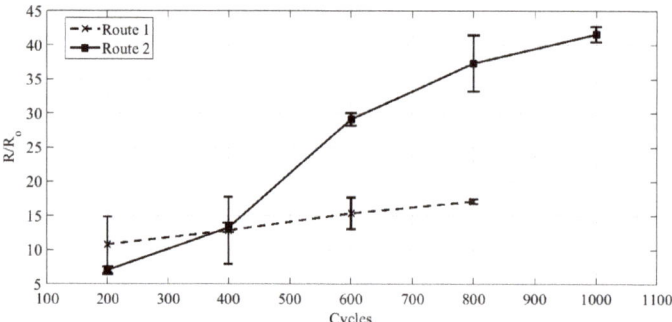

Figure 9. Reliability curves of expanded graphite thin films. Error bars show standard deviations between the tested samples.

4. Conclusions

To conclude, this paper presented the mechanical stretchability, flexibility and reliability of expanded graphite thin films on polymer substrate PDMS. The expanded graphite particles were prepared using the bath sonication of commercial graphite flakes (the first route) and graphite particles obtained through electrolysis (the second route) under interstitial substitution. It was concluded that, due to bath sonication and favorable effects of electrolysis and sulfate ions, the intercalation was much more effective, resulting in lower initial film resistances. Additionally, from the perspective of mechanical characterizations, the films prepared by following the second route seem to over perform the films prepared by the first route. However, due to the strong dependency on the piezoresistive part, a higher change in resistance was observed during the mechanical loading of the films prepared by the second route. Nevertheless, both types of film were found to remain stretchable, flexible, and functional, even at repeated flexures. Various results in this paper have shown that the as-fabricated expanded graphite thin films demonstrate favorable mechanical performance, which can be used in various stretchable and flexible electronics applications.

Author Contributions: M.M.N. did the conceptualization, methodology and funding acquisition; M.M. designed the experiments and wrote the manuscript; H.A. (Hasan Askari), B.A.B., A.U.R., H.A. (Hassan Abbas), Z.H. and D.Z. performed the experiments; D.H., J.H.Z. and M.S.A.B. did the analyses and edited the manuscript. All authors have read and agreed to the published version of the manuscript.

Funding: This research was funded by the Universiti Brunei Darussalam's University Research Grant Number UBD/RSCH/URC/RG(b)/2019/008.

Acknowledgments: The authors would like to thank NED University of Engineering and Technology Pakistan for its support.

Conflicts of Interest: The authors declare no conflict of interests.

References

1. Wu, Y.; Pan, Q.; Zheng, F.; Ou, X.; Yang, C.; Xiong, X.; Liu, M.; Hu, D.; Huang, C. Sb@C/expanded graphite as high-performance anode material for lithium ion batteries. *J. Alloy Compd.* **2018**, *744*, 481–486. [CrossRef]
2. Kim, J.; Yoon, G.; Kim, J.; Yoon, H.; Baek, J.; Lee, J.H.; Kang, K.; Jeon, S. Extremely large, non-oxidized graphene flakes based on spontaneous solvent insertion into graphite intercalation compound. *Carbon* **2018**, *139*, 309–316. [CrossRef]
3. Lee, S.W.; Lee, W.; Hong, Y.; Lee, G.; Yoon, D.S. Recent advances in carbon material-based NO_2 gas sensors. *Sens. Actuators B Chem.* **2018**, *255*, 1788–1804. [CrossRef]

4. Ali, K.; Choi, K.-H.; Muhammad, N.M. Roll-to-roll atmospheric atomic layer deposition of Al_2O_3 thin films on PET substrates. *Chem. Vap. Dep.* **2014**, *20*, 380–387. [CrossRef]
5. Li, C.; Xie, B.; Chen, D.; Chen, J.; Li, W.; Chen, Z.; Gibb, S.W.; Long, Y. Ultrathin graphite sheets stabilized stearic acid as a composite phase change material for thermal energy storage. *Energy* **2019**, *166*, 246–255. [CrossRef]
6. Dao, V.-D.; Choi, H.-S. Carbon-Based Sunlight Absorbers in Solar-Driven Steam Generation Devices. *Glob. Chall.* **2008**, *2*, 1700094. [CrossRef]
7. Natarajan, S.; Lakshmi, D.S.; Bajaj, H.C.; Srivastava, D.N. Recovery and utilization of graphite and polymer materials from spent lithium-ion batteries for synthesizing polymer-graphite nanocomposite thin films. *J. Environ. Chem. Eng.* **2015**, *3*, 2538–2545. [CrossRef]
8. Afify, A.S.; Ahmad, S.; Khushnood, R.A.; Jagdale, P.; Tulliani, J.-M. Elaboration and Characterization of novel humidity sensor based on micro-carbonized bamboo particles. *Sens. Actuators B Chem.* **2017**, *239*, 1251–1256. [CrossRef]
9. Jeong, J.; Kim, S.; Cho, J.; Hong, Y. Stable Stretchable Silver Electrode Directly Deposited on Wavy Elastomeric Substrate. *IEEE Electron Device Lett.* **2009**, *30*, 1284–1286. [CrossRef]
10. Kim, D.-H.; Lu, N.; Ma, R.; Kim, Y.-S.; Kim, R.-H.; Wang, S.; Wu, J.; Won, S.M.; Tao, H.; Islam, A.; et al. Epidermal electronics. *Science* **2011**, *333*, 838–843. [CrossRef]
11. Lipomi, D.J.; Tee, B.C.-K.; Vosgueritchian, M.; Bao, Z. Stretchable organic solar cells. *Adv. Mater.* **2011**, *23*, 1771–1775. [CrossRef] [PubMed]
12. White, M.S.; Kaltenbrunner, M.; Głowacki, E.D.; Gutnichenko, K.; Kettlgruber, G.; Graz, I.; Aazou, S.; Ulbricht, C.; Egbe, D.A.M.; Miron, M.C.Z.; et al. Ultrathin, highly flexible and stretchable PLEDs. *Nat. Photonics* **2013**, *7*, 811–816. [CrossRef]
13. Brown, E.W. Octacalcium phosphate and hydroxyapatite. *Nature* **1962**, *196*, 1048–1050. [CrossRef]
14. Cho, C.-K.; Hwang, W.-J.; Eun, K.; Choa, S.-H.; Na, S.-I.; Kim, H.-K. Mechanical flexibility of transparent PEDOT:PSS electrodes prepared by gravure printing for flexible organic solar cells. *Sol. Energy Mater. Sol. Cells* **2011**, *95*, 3269–3275. [CrossRef]
15. Lee, J.-H.; Lee, K.Y.; Gupta, M.K.; Kim, T.Y.; Lee, D.-Y.; Oh, J.; Ryu, C.; Yoo, W.J.; Kang, C.-Y.; Yoon, S.-J.; et al. Highly stretchable piezoelectric-pyroelectric hybrid nanogenerator. *Adv. Mater.* **2007**, *26*, 765–769. [CrossRef]
16. Geim, A.K.; Novoselov, K.S. The rise of graphene. *Nat. Mater.* **2007**, *6*, 183–191. [CrossRef]
17. Nuvoli, D.; Valentini, L.; Alzari, V.; Scognamillo, S.; Bon, S.B.; Piccinini, M.; Illescas, J.; Mariani, A. High concentration few-layer graphene sheets obtained by liquid phase exfoliation of graphite in ionic liquid. *J. Mater. Chem.* **2011**, *21*, 3428–3431. [CrossRef]
18. Niu, L.; Li, M.; Tao, X.; Xie, Z.; Zhou, X.; Raju, A.P.A.; Young, R.J.; Zheng, Z. Salt-assisted direct exfoliation of graphite into high-quality, large-size, few-layer graphene sheets. *Nanoscale* **2013**, *5*, 7202–7208. [CrossRef]
19. Jin, J.; Leesirisan, S.; Song, M. Electrical conductivity of ion-doped graphite/polyethersulphone composites. *Compos. Sci. Technol.* **2010**, *70*, 1544–1549. [CrossRef]
20. Parvez, K.; Wu, Z.S.; Li, R.; Liu, X.; Graf, R.; Feng, X.; Müllen, K. Exfoliation of graphite into graphene in aqueous solutions of inorganic salts. *J. Am. Chem. Soc.* **2014**, *136*, 6083–6091. [CrossRef]
21. Kato, R.; Hasegawa, M. Fast synthesis of thin graphite film with high-performance thermal and electrical properties grown by plasma CVD using polycrystalline nickel foil at low temperature. *Carbon* **2019**, *141*, 768–773. [CrossRef]
22. Yudasaka, M.; Kikuchi, R.; Matsui, T.; Kamo, H.; Ohki, Y.; Yoshimura, S.; Ota, E. Graphite thin film formation by chemical vapor deposition. *Appl. Phys. Lett.* **1994**, *64*, 842–844. [CrossRef]
23. Luo, S.; Liu, T. SWCNT/graphite nanoplatelet hybrid thin films for self-temperature- compensated, highly sensitive, and extensible piezoresistive sensors. *Adv. Mater.* **2013**, *25*, 5650–5657. [CrossRef] [PubMed]
24. Lang, U.; Rust, P.; Schoberle, B.; Dual, J. Piezoresistive properties of PEDOT:PSS. *Microelectron. Eng.* **2009**, *86*, 330–334. [CrossRef]
25. Politano, A.; Chiarello, G. Probing the Young's modulus and Poisson's ratio in graphene/metal interfaces and graphite: A comparative study. *Nano Res.* **2015**, *8*, 1847–1856. [CrossRef]
26. Suo, Z.; Ma, E.Y.; Gleskova, H.; Wagner, S. Mechanics of rollable and foldable film-on-foil electronics. *Appl. Phys. Lett.* **1999**, *74*, 1177–1179. [CrossRef]

© 2020 by the authors. Licensee MDPI, Basel, Switzerland. This article is an open access article distributed under the terms and conditions of the Creative Commons Attribution (CC BY) license (http://creativecommons.org/licenses/by/4.0/).

Article

Optimization of Synthesizing Upright ZnO Rod Arrays with Large Diameters through Response Surface Methodology

Xiaofei Sheng [1,2], Yajuan Cheng [3,*], Yingming Yao [3,*] and Zhe Zhao [4,5,*]

1. School of Materials Science and Engineering, Hubei University of Automotive Technology, Shiyan 442002, China; auden1@126.com
2. School of Mechanical and Automotive Engineering, Zhejiang University of Water Resources and Electric Power, Hangzhou 310018, China
3. Key Laboratory of Organic Synthesis of Jiangsu Province and the State and Local Joint Engineering Laboratory for Novel Functional Polymeric Materials, College of Chemistry, Chemical Engineering and Materials Science, Soochow University, Suzhou 215123, China
4. Department of Materials Science and Engineering, KTH Royal Institute of Technology, SE-100 44 Stockholm, Sweden
5. Department of Materials Science and Engineering, Shanghai Institute of Technology, Shanghai 201418, China
* Correspondence: yjcheng@suda.edu.cn (Y.C.); yaoym@suda.edu.cn (Y.Y.); zhezhao@kth.se (Z.Z.)

Received: 16 April 2020; Accepted: 29 May 2020; Published: 31 May 2020

Abstract: The deposition parameters involved in chemical bath deposition were optimized by a response surface methodology to synthesize upright ZnO rod arrays with large diameters. The effects of the factors on the preferential orientation, aspect ratio, and diameter were determined systematically and efficiently. The results demonstrated that an increased concentration, elevated reaction temperature, prolonged reaction time, and reduced molar ratio of Zn^{2+} to tri-sodium citrate could increase the diameter and promote the preferential oriented growth along the [002] direction. With the optimized parameters, the ZnO rods were grown almost perfectly vertically with the texture coefficient of 99.62. In the meanwhile, the largest diameter could reach 1.77 μm. The obtained rods were merged together on this condition, and a dense ZnO thin film was formed.

Keywords: tri-sodium citrate; ZnO rod arrays; response surface methodology

1. Introduction

The application of nanomaterials has been extended to many fields with the development of new synthesis methods [1,2] and has brought many exciting achievements [3]. Vertically aligned ZnO rod arrays have attracted more and more research interest over the past few years because of their remarkable exhibition or unique properties and exciting potential applications, such as solar cells [4–6], piezoelectric nanogenerators [7,8], light emitting diodes [9,10], thermoelectrics [11–15], and sensors [16,17]. With tremendous applications for vertically aligned ZnO rod arrays, searching a method to control their morphologies and properties is crucial to optimize the arrays for specific tasks. In order to satisfy the demands of various applications, it is necessary to both understand the growth mechanisms of the vertically aligned ZnO rod arrays and to find a method to control the properties with the guidance of those mechanisms. In particular, certain applications require thin rods with high aspect ratios, while others necessitate thick rods with low aspect ratios. Moreover, if the rods are perfectly upright to the substrate and the rods are coarsened, the rods may merge together to form a dense ZnO thin film.

Based on the above consideration, efforts were taken to increase the diameter of the obtained ZnO rods. Moreover, the aspect ratios were also lowered by increasing the diameters. In our previous studies [18,19], ZnO rod arrays with high aspect ratios were successfully synthesized. By combining these works, the aspect ratio can be systematically and precisely controlled to meet various demands. Tri-sodium citrate has been reported as an effective surface modifier to promote the lateral growth of ZnO rods [20–24]. Das et al. explored ZnO morphologies' evolution with increased citrate concentrations [20] and found that growth with excess citrate obtained spherical particles with very low crystallinity instead of rods or plates. In this work, the boundary value of the citrate concentration to get ZnO plates in Das's work was used as the highest value of the citrate concentration, and the impact of the citrate concentration on the diameter was studied systematically in a low range. Chemical bath deposition was employed to synthesize ZnO rods arrays in this work due to its cost-efficiency, simple process, and ease to scale-up. In solution growth, the changing of other growth parameters (precursor concentration, reaction temperature, and reaction time) also contributes to the increase of the diameters. Therefore, the influences of these parameters were also investigated. Since various precursor concentrations were employed, the molar ratio of the zinc ion to the tri-sodium citrate instead of the citrate concentration was used in this work to more accurately study the impact of the citrate.

In this study, four factors, including the precursor concentration, reaction temperature, reaction time, and the molar ratio of zinc ion to the citrate, were investigated. If the conventional 'changing one separate factor at a time' (COST) approach was applied to optimize the parameters, plenty of experiments along with a large amount of time and money would be required. Moreover, the interactions between the factors could not be investigated, and the optimal settings of factors would be possible to miss. Therefore, it was essential to find an approach to efficiently identify the optimal parameters and to comprehensively and thoroughly guide the investigation. As a result, a powerful statistical method called the response surface methodology was employed in this work. Through this technique, multiple process variables could be varied simultaneously, and only a few experimental trials were required to identify the influence of each factor. Additionally, the interactions between variables could also be identified and quantified. Moreover, a mathematical model could be proposed to describe the synthesis processes and predict the response with specific parameters. Furthermore, a response surface plot and a 4D contour plot as a function of the independent parameters could be obtained to determine the optimal points.

In this work, vertically aligned ZnO rods arrays with large diameters were successfully synthesized. The influence of each factor on the growth quality was investigated through a comprehensive study. With the assistance of response surface methodology in this work, maps of the changing tendency were also created to predict the growth quality of the ZnO rod arrays.

2. Experimental Sections

2.1. The Pre-Treatment of the Substrates

Small pieces C-plane (0001) sapphire (Guangdong Orient Zirconic Ind Sci & Tech Co., Ltd., Shantou, China) with dimensions of 12.5 × 12.5 mm were used as the substrates in this work. A cleaning sequence developed by Kern et al. [25] was employed to pretreat the substrates. This aimed to ensure that the surface of the substrates was completely clean and hydrophilic, and then seeds could be homogenously coated onto the whole substrates during the next step. The cleaning sequence included four steps with, intermediate rinsing steps separating each chemical step. Firstly, the substrates were ultrasonicated in ethanol for 3 min to remove some of the organic contamination. Subsequently, the substrates were heated in a piranha solution (1:3, 30% H_2O_2/H_2SO_4) (Sigma–Aldrich China, Shanghai, China) at 120 °C for 30 min to remove relatively heavy organic contaminations. Thirdly, the substrates were immersed into an SC-1 solution (1:1:5 25% NH_4OH/30% H_2O_2/Milli-Q H_2O) (Sigma–Aldrich China, Shanghai, China) at 75 °C for 15 min to remove particles and metals. Finally, an SC-2 solution (1:1:6 35% HCl, VWR/30% H_2O_2/Milli-Q H_2O) (Sigma–Aldrich China, Shanghai, China) was employed to remove the residual metals, including metals that may have been deposited in the SC-1 solution. The substrates were immersed in the SC-2 solution and heated at 75 °C for 15 min. The substrates were flash-air dried at room temperature before the seeding step.

2.2. Deposition of Seed Layers on the Substrates

The preparation procedure was done in the same manner as reported in our previous study [18]. Firstly, a droplet of a 40 μL solution of 0.01 M zinc acetate dihydrate (98%, Sigma–Aldrich China, Shanghai, China) ethanol solution was uniformly spin-coated on the pre-cleaned substrates. The spin coating parameters were set as: v = 400 rpm and t = 60 s. After the substrate was dried, the coating was repeated two additional times. This aimed to ensure that a complete and uniform layer of zinc acetate crystallites was obtained on all of the substrates. Subsequently, the coated substrates were annealed in a furnace at 500 °C for 2 h. Through the decomposition of the zinc acetate, a zinc oxide layer was yielded.

2.3. Pre-Growth of ZnO Nanorods

ZnO nanorods were grown by suspending the seeded substrates upside-down in a sealed bottle (as shown in Figure 1). In the bottle, 30 mL of a 0.05 M zinc nitrate hydrate solution (Shanghai Aladdin Biochemical Technology Co., Ltd., Shanghai, China) and 30 mL of a 0.05 M hexamethylenetetramine (HMT) aqueous solution (Shanghai Aladdin Biochemical Technology Co., Ltd., Shanghai, China) were mixed homogenously. The bottle was heated at 90 °C for 2 h. After it was cooled down, the substrates were removed and cleaned with ethanol.

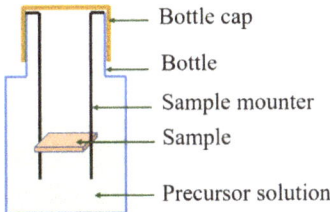

Figure 1. The scheme of the reactor.

2.4. Preparation of Dense ZnO Films

After the pre-growth of the ZnO nanorods, a fresh equimolar mixed solution consisting of 30 mL of zinc nitrate hydrate and 30 mL of hexamethylenetetramine (HMT) was filled into the bottles. To promote the growth along the lateral direction, a specific amount of tri-sodium citrate (Shanghai Aladdin Biochemical Technology Co., Ltd., Shanghai, China) was added into the mixed solution. Then, the substrates were mounted upside down in the bottle. Subsequently, the bottle was capped tightly and heated at a specific temperature for a certain duration. After the bottle was cooled down naturally, the samples were taken out and rinsed by ethanol.

2.5. Characterization of the Obtained ZnO Rods Arrays

The crystal structure and orientation of the obtained samples were analyzed by a powder X-ray diffractometer (XRD, PANalytical powder X-ray diffractometer with a Cu Kα1 radiation, λ = 0.15406 nm) (PANalytical, Holland). In order to characterize the degree of preferential orientation of the nanorods along the (002) plane, the relative texture coefficient (denoted as TC_{002}) was calculated for each sample. This was the ratio of diffraction peaks (002) over (100) and (101) in XRD patterns, and it was calculated as follows:

$$TC_{002} = \frac{I_{002}/I^0_{002}}{I_{100}/I^0_{100} + I_{002}/I^0_{002} + I_{101}/I^0_{101}} \quad (1)$$

where I_{100}, I_{002}, and I_{101} are the measured diffraction intensities of the (100), (002), and (101) planes, respectively. I_{100}^0, I_{002}^0, and I_{101}^0 are the corresponding values of the standard data (JCPDS 00-036-1451).

The morphologies of the obtained samples were explored by using a field emission scanning electron microscope (JEOL, JSM-7000F, Japan). The mean diameter of the obtained ZnO rods was determined as an average of ten measured rods of the top view of the SEM images. The aspect ratio was calculated as the ratio of the rod length to the mean diameter. Photoluminescence was tested by a Photoluminescence spectrophotometer (Varian CaryEclipse, US). The spectra were examined by using a He–Cd laser with an excitation wavelength of 279 nm at room temperature.

2.6. Experimental Design

The design of the experiments and the analysis of the obtained results were guided by the commercial software MODDE (version 10, Umetrics, Umeå, Sweden, 2015). The response surface methodology (RSM) was employed to build models, to assess the effect of factors, and to identify the optimal conditions. The significance and adequacy of the model was tested by ANOVA. Additionally, response surface plots and 4D contour plots were also created to find the optimal parameters. Moreover, a one-pair-at-a-time main effect analysis was also applied to investigate the impact of the factors on the responses.

3. Results and Discussion

3.1. Optimization of the Experimental Parameters

In this work, the following four experimental factors were studied: (i) the concentration of Zn^{2+} (c), (ii) reaction temperature (T), (iii) reaction time (t), and (iv) the molar ratio of Zn^{2+} to tri-sodium citrate (R). A central composite face (CCF) design composed of a full factorial design and star points placed on the faces of the sides was employed in this stage. The detailed parameters of the design and the achieved results are listed in Table 1. Obviously, most of achieved texture coefficients were above 90, indicating a good oriented growth along the [0002] direction. Most of the aspect ratios were limited in the range of 1–3, which was largely decreased compared to the results reported in our previous work [18]. Moreover, the values of diameters were varied in a broad range, which would meet various demands for different applications [26,27].

The outputs of the responses were analyzed by the MODDE software, and a model was built. Several evaluating tools were employed to investigate the adequacy of the model and to determine the changing tendency of the response with the factors. Firstly, the ANOVA was conducted to evaluate the adequacy of the model. The tabulated output is presented in Table 2. As mentioned in our previous study [28], a regression model is significant and has no lack of fit if the first p value is smaller than 0.05. Therefore, the model for the diameter was valid and adequate, while the other two models were insignificant. This was because that most values of the obtained texture coefficients and most of aspect ratios were limited in a narrow range, while only a few values were scattered away from the region. Thus, the effect of the reaction parameters was not significant enough to be detected in the selected range of the parameters. In other words, good alignment could be obtained in most cases within the selected ranges in Table 1. In our previous work [18], ZnO rods with texture coefficient higher than 0.95 were always obtained on condition of c = 0.06 – 0.1 M, T ≥ 70 °C, and t ≥ 6 h. In comparison, there was a slight decrease for most values of texture coefficient in this work, and several values were even far away from 90%. This suggests that the tri-sodium citrate may have had a slight effect on the preferential oriented growth. With respect to the aspect ratio, most of the values were limited in a narrow range 1–3 μm, which was reduced to a large extent compared to the previous results [18]. This reflects that the aspect ratio could be efficiently controlled by the addition of tri-sodium citrate.

Table 1. Optimization design with detailed factors and the obtained responses.

No.	c/M	T/°C	t/h	R/	TC$_{002}$/	Aspect Ratio/	D/µm
1	0.06	70	6	500	80.71	6.61	0.26
2	0.1	70	6	500	95.92	4.76	0.36
3	0.06	90	6	500	96.12	1.45	0.70
4	0.1	90	6	500	69.05	1.45	1.20
5	0.06	70	16	500	91.84	1.23	1.47
6	0.1	70	16	500	97.67	1.71	1.03
7	0.06	90	16	500	97.95	1.31	1.11
8	0.1	90	16	500	99.62	0.96	1.77
9	0.06	70	6	1500	71.27	1.57	0.71
10	0.1	70	6	1500	95.09	2.51	0.52
11	0.06	90	6	1500	92.53	9.28	0.29
12	0.1	90	6	1500	97.90	2.55	0.71
13	0.06	70	16	1500	98.58	2.20	0.73
14	0.1	70	16	1500	91.27	2.30	0.75
15	0.06	90	16	1500	98.89	7.92	0.31
16	0.1	90	16	1500	97.04	2.42	0.71
17	0.06	80	11	1000	98.24	2.20	0.80
18	0.1	80	11	1000	94.53	2.18	0.96
19	0.08	70	11	1000	83.39	1.63	0.99
20	0.08	90	11	1000	98.35	1.95	0.74
21	0.08	80	6	1000	98.41	4.78	0.43
22	0.08	80	16	1000	98.43	1.62	1.08
23	0.08	80	11	500	94.76	2.30	1.20
24	0.08	80	11	1500	99.67	3.00	0.66
25	0.08	80	11	1000	93.11	1.65	1.18
26	0.08	80	11	1000	95.42	1.84	1.12
27	0.08	80	11	1000	93.10	1.78	1.27

Since the model of diameters was valid, a fitted regression equation was established by using partial least squares projections to investigate the relationship between the diameter and the independent parameters:

$$\begin{aligned}\log_{10}(D+0.5) = &-0.558091 - 17.9492*c - 0.000649181*T + 0.155156*t \\ &+0.000987155*R - 0.0032472*t^2 + 0.244749*c*T \\ &-0.000618995*T*t - 1.10566e^{-5}*T*R - 2.02695e^{-5}*t*R\end{aligned} \quad (2)$$

where D presents diameter, c is the concentration of the precursor, T is the reaction temperature, t is the reaction time, and R is the molar ratio of Zn^{2+} to tri-sodium citrate.

Figure 2 shows the plot of the observed against predicted diameters. More specifically, the observed values were from the experimental diameters, while the predicted ones were calculated from the above equation. It was obvious that almost all of the observed values were very close to the predicted ones in Figure 2. This means that the model was adequate and it had a good prediction power. Moreover, as the R^2 value 0.89 was close to the ideal value 1, the regression model fit the raw data well. Therefore, the model of diameter could be regarded to be an adequate model for the prediction of diameters.

Table 2. The ANOVA table of the three responses.

TC_{002}	DF [a]	SS [b]	MS [c]	F	p
Total	27	236,668	8765.49		
Constant	1	234,987	234,987		
Total corrected	26	1681.06	64.656		
Regression	6	764.311	127.385	2.779	0.039
Residual	20	916.751	45.838		
Lack of Fit (Model error)	18	913.178	50.732	28.399	0.035
Pure error (Replicate error)	2	3.573	1.786		
Aspect Ratio	**DF [a]**	**SS [b]**	**MS [c]**	**F**	**p**
Total	27	5.474	0.203		
Constant	1	4.001	4.001		
Total corrected	26	1.472	0.057		
Regression	4	0.397	0.099	2.032	0.125
Residual	22	1.075	0.049		
Lack of Fit (Model error)	20	1.074	0.054	106.606	0.009
Pure error (replicate error)	2	0.001	0.001		
Diameter	**DF [a]**	**SS [b]**	**MS [c]**	**F**	**p**
Total	27	1.611	0.060		
Constant	1	0.364	0.364		
Total corrected	26	1.247	0.048		
Regression	9	1.104	0.123	14.541	0.000
Residual	17	0.143	0.008		
Lack of Fit (Model error)	15	0.142	0.009	12.629	0.076
Pure error (replicate error)	2	0.001	7.4×10^{-4}		

[a] degree of freedom, [b] sum of squares, and [c] means square.

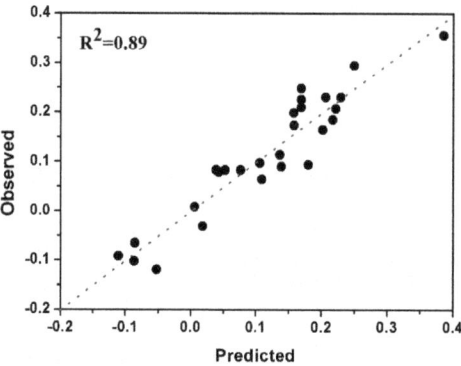

Figure 2. The observed versus predicted values of the diameters.

A scaled and centered coefficients plot, as shown in Figure 3, was applied to investigate the effects of the parameters. If the confidence interval of the factor did not cross the horizontal axis, the corresponding factor could be determined as a significant parameter for the response. To illustrate, the linear terms (c, t, and R), the square term t^2, and the interaction terms (c*T, T*t, T*R, and t*R) were identified as significant parameters for the diameters. In addition, the diameters is increased with an increased concentration (c), reaction time (t), and interactions between the concentration and reaction temperature (c*T). On the other hand, the increase of the molar ratio (R), the quadratic of reaction time (t^2), and the interaction terms T*t, T*R, and t*R decreased the diameters. What should be noted is that there was one optimal point for the reaction time to achieve the largest diameter due to the negative effect of t^2 and the positive effect of t.

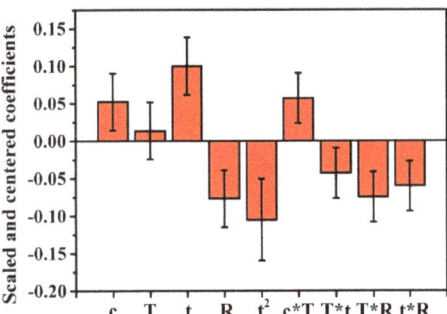

Figure 3. The scaled and centered coefficients plot of the diameters.

To further understand the changing tendency of the response with the parameters, the response surface plots and response 4D contour plot are depicted. In the response surface plot, two parameters were used as two axes, while the other two parameters were kept at their middle levels. Figure 4 shows the response surface plots of the diameters. Since the interaction terms c*T, T*t, T*R, and t*R were detected in Figure 3, the response surface plots using them as axes are depicted in Figure 4a–d, respectively. It can be seen from Figure 4a that an increased precursor concentration and elevated reaction temperature were advantageous to the coarsening of the rods. The curved surface in Figure 4b shows that there was an optimal region at 11–16 h for the reaction time to achieve the biggest diameters. This was in a good agreement with the conclusion obtained from the scaled and centered coefficients plot. Both of the surfaces in Figure 4c,d show that diameters were increased with a reduced molar ratio, which certified the conclusion achieved in Figure 3. Additionally, Figure 4d shows that the biggest diameter could always be obtained in an optimal region of 11–16 h for the reaction time. This certificated the existence of the optimal region for reaction time again.

Figure 5 displays the response 4D contour plot of the diameters, which gives an overall view of the changing tendency of the response. Similar to the conclusion from Figure 3, an optimal region for reaction time to achieve the largest diameter could also be detected but with a narrow range of 11–13 h. The comparison among the three columns certified again that the diameter increases with an increased concentration. Moreover, the interactions between the factors could also be clearly observed from the curved lines in Figure 5. Apparently, when both the concentration and reaction temperature were set as their high level, ZnO rods with large diameters could always be obtained. This was consistent with the results obtained from Figure 4a. Furthermore, the required process parameters for the specific diameter could be easily found, which made the practical process efficient.

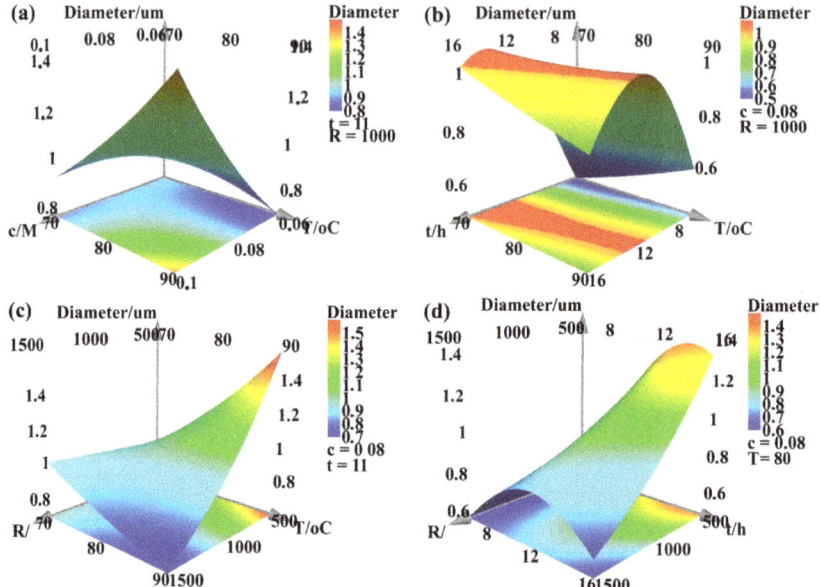

Figure 4. The response surface plots of the diameters with varied parameters while the other two parameters were fixed: (**a**) varied the concentration of Zn^{2+} (c) and reaction temperature (T), while reaction time (t) = 11h and the molar ratio of Zn^{2+} to tri-sodium citrate (R) = 1000; (**b**) varied t and T while c = 0.08 M and R = 1000; (**c**) varied R and T while c = 0.08 M and t = 11 h; and (**d**) varied R and t, while c = 0.08 M and T = 80 °C.

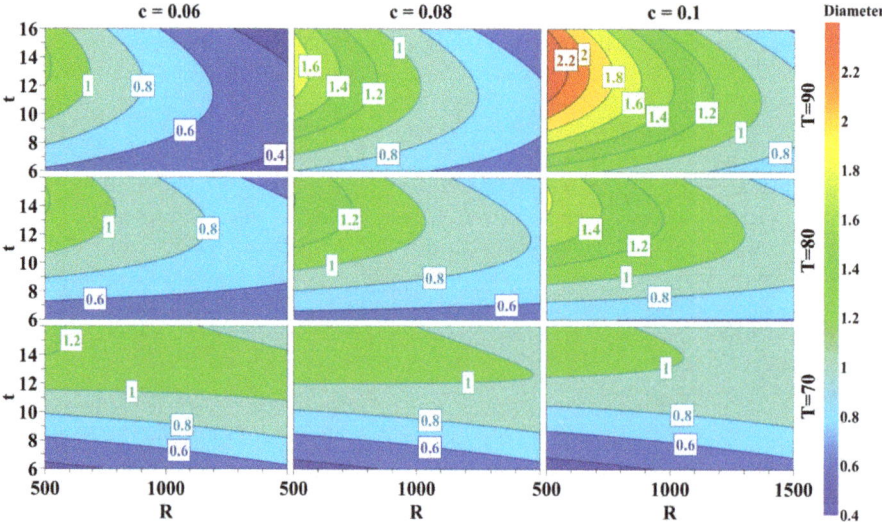

Figure 5. The response 4D contour plot of the diameters. The three columns of images represent the changing tendency of diameter with varied reaction temperatures and molar ratios of Zn^{2+} to tri-sodium citrate when c = 0.06, 0.08, and 0.1 M, and the three rows of images represent the changing tendency of diameter with varied reaction temperatures and molar ratios of Zn^{2+} to tri-sodium citrate when T = 70, 80, and 90 °C.

Since the models of the texture coefficients and aspect rations had a lack of fit, they could not be fitted by MODDE. On this condition, a one-pair-at-a-time main effect analysis was instead applied to identify good factor settings. With this method, the experiments were grouped into pairs where only one factor was varied. The first sixteen experiments were selected to illustrate this method, and the plots for texture coefficients and aspect ratios are shown in Figures 6 and 7, respectively. In these two figures, it is evident that the changing tendency of the response with the factors was complicated. However, the general tendency could still be identified. For instance, in Figure 6, the texture coefficient can be seen to have increased with an increased concentration, an elevated reaction temperature, and a prolonged reaction time in most cases. The preference for molar ratio was not obvious. Moreover, the texture coefficient could always reach the value larger than 90% when the reaction temperature was set to 16 h. In the meantime, the value of texture coefficient was always below 85% when the parameters were set as c = 0.06 M, T = 70 °C, and t = 6 h. In terms of aspect ratio, there was no obvious preference for precursor concentration and reaction temperature. However, the decrease of reaction time and increase of molar ratio favored the increase of aspect ratio in most cases. Furthermore, aspect ratio higher than 8 could always be obtained with the parameters c = 0.06 M, T = 80 °C, and R = 1500.

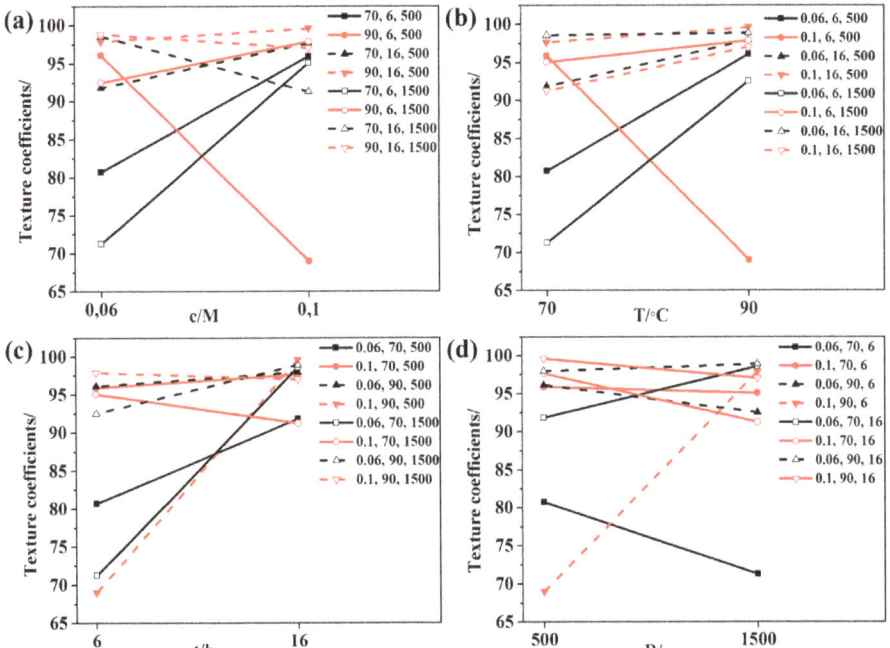

Figure 6. One-pair-at-a-time main effect analysis of varied parameters (**a**) the concentration of Zn^{2+}, (**b**) temperature, (**c**) reaction time, and (**d**) molar ration of Zn^{2+} to tri-sodium citrate on the texture coefficients for the first sixteen experiments in Table 1.

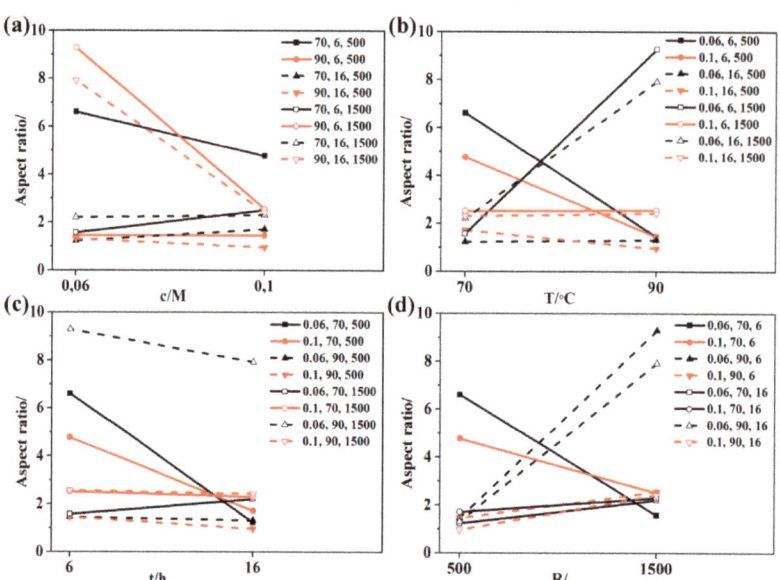

Figure 7. One-pair-at-a-time main effect analysis of varied parameters (**a**) the concentration of Zn^{2+}, (**b**) temperature, (**c**) reaction time, and (**d**) molar ration of Zn^{2+} to tri-sodium citrate on aspect ratios for the first sixteen experiments in Table 1.

In summary, the preferential growth along the [0002] direction was favored when the concentration, reaction temperature, and reaction time were set at the high level while the molar ratio was set at the low level. Moreover, the aspect ratio was increased by shortening the reaction time and increasing the molar ratio. Furthermore, the diameters of the obtained ZnO rods were increased with the increased concentration and reduced molar ratio. There was an optimal region of reaction time of 11–13 h to get the largest diameter. On the condition of high reaction temperature and high precursor concentration, a large diameter could always be obtained. In combining the above results to achieve ZnO rods with high preferential orientation and big diameter, the precursor concentration, reaction temperature, reaction time, and molar ratio should be set as 0.1 M, 90 °C, 16 h, and 500, respectively. The optimal condition was exact the experimental setup of sample No. 8, where both the texture coefficient and the diameter got the largest values.

The mechanism for the effect of the significant factors including concentration, reaction time, and molar ratio can be elucidated as follows: Due to the addition of tri-sodium citrate, the growth rate of the (0001) facets was suppressed. The narrow distribution of the length of 1–3 µm directly reflected this suppression. On this condition, the increase of the concentration promoted the thickening of the rods. Similarly, when R was reduced, the amount of the tri-sodium citrate was correspondingly increased. Consequently, the suppression of the growth along the [0001] direction was stronger and the diameter was increased. Compared to the obtained diameters of 100–150 nm in our previous study [18] without the addition of tri-sodium citrate, the diameter was increased to a large extent in this work. With respect to the reaction time, the rods got thickening before 12 h. This was reasonable because a prolonged reaction time was favorable for the lateral growth. However, the slight reduction after 12 h is difficult to explain. More investigation needs to carry out.

3.2. Structure and Morphologies of the Obtained ZnO Rod Arrays

The synthetic approach in this work enabled the production of dense microstructures for all of samples. The morphologies evolved with an increased diameter. Figure 8a–d shows the typical

SEM images of these structures. For all the samples, since the ZnO seeds may have been spanned off at the edge of the sample and rod arrays were formed through spatially confined oriented growth, the rods on edges were randomly oriented. The corresponding SEM images are shown in Figure S1 in the Supplemental Information. However, the obtained ZnO rods were vertically and homogenously aligned in the center. Therefore, all the SEM images in this work were taken in the center of each sample, and the lengths and diameters were averaged over 20 randomly picked rods. Apparently, the obtained structures were denser with an increased diameter. In Figure 8a, the mean diameter can be seen to be 0.29 µm, and the rods are individual. When the mean diameter was increased to 0.43 µm in Figure 8b, the rods started to merge with each other. In Figure 8c, the diameter was further increased to 0.66 µm, all the rods were merged into each other and a dense thin film was formed according to the corresponding cross-sectional SEM images. However, further increasing the diameter caused secondary growth, as shown in Figure 8d. Layered structures were produced on top of the rod arrays on this condition. The corresponding XRD patterns of the samples are also shown in Figure 8e for the investigation the crystal structures. It is evident that all of the samples had a dominated preferential orientation along the [002] direction. More specifically, for samples 24 and 8, the (100) and (101) peaks were almost invisible, and perfectly (002) oriented growth was obtained. This was consistent with the morphologies observed from the corresponding cross sectional SEM images in Figure 8c,d. In contrast, the observation of the (100) and (101) peaks in the patterns of samples 11 and 21 indicated the slight inclining of the obtained rods.

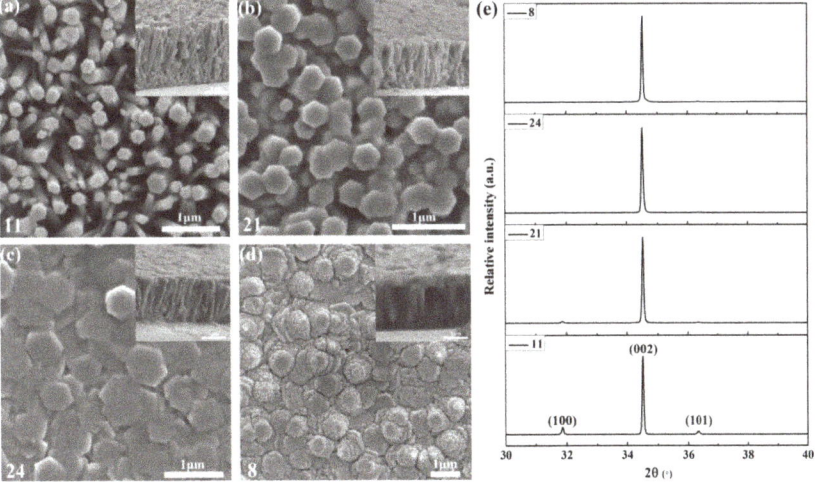

Figure 8. Typical SEM images (**a**–**d**) and XRD patterns (**e**) of the obtained ZnO rod arrays. The insets in the SEM images are the corresponding cross-section images of ZnO nanorod arrays, and the enlarged cross-sectional images can be found in Figure S2 in the Supplemental Information. The numbers denoted in the plots are the sample number in Table 1.

3.3. Photoluminescence Spectra of the Obtained ZnO Rod Arrays

The room temperature PL spectra of the obtained ZnO rods were measured to investigate the optical properties of the obtained ZnO rod arrays. If the spectra of all the samples are depicted and discussed here, the paper would be too long. Therefore, only the PL spectra of the four samples selected in Figure 8 are presented here, as shown in Figure 9. There were two peaks in the spectra. One was a strong near-band edge UV emission peak located at ~379 nm which originated from the recombination of the electrons from conduction bands to valence bands. The other was a broad visible band ranging between 500 and 650 nm. This was a defect related deep level emission induced by the defects of

O vacancies, Zn interstitials, or their complexes. It was evident that the photoluminescence properties of the obtained ZnO rod arrays were related to the texture coefficient and the diameter. Apparently, the higher the texture coefficient, the lower the intensity of the broad-band emission. This indicated that the sample with higher texture coefficient had fewer defects and better crystal quality. Comparing the spectra of sample 8 and 24, which had almost the same texture coefficient, the crystal quality was better when the diameter was larger. This can be explained as follows: ZnO rod arrays were obtained by spatially confined oriented growth (as explained in our previous work [19]), so less random orientation close to the substrate appeared when the diameter was larger. Therefore, fewer boundaries were obtained when the rods merged together and the crystal quality was better. What should be noted is that the decrease of the intensity of the broad visible band may also have been related to the location of defects. It has been suggested that defects that are optically active are situated at the near surface of ZnO nanowires [29], while the NBE peak could be attributed to the whole volume of the nanowire. Therefore, larger diameter rods had a smaller surface area to volume ratio that could have caused the reduced DLE peak relative to the NBE peak. Both the above two mechanisms were related to the fewer surfaces of rods with larger diameters, i.e., increased crystal qualities.

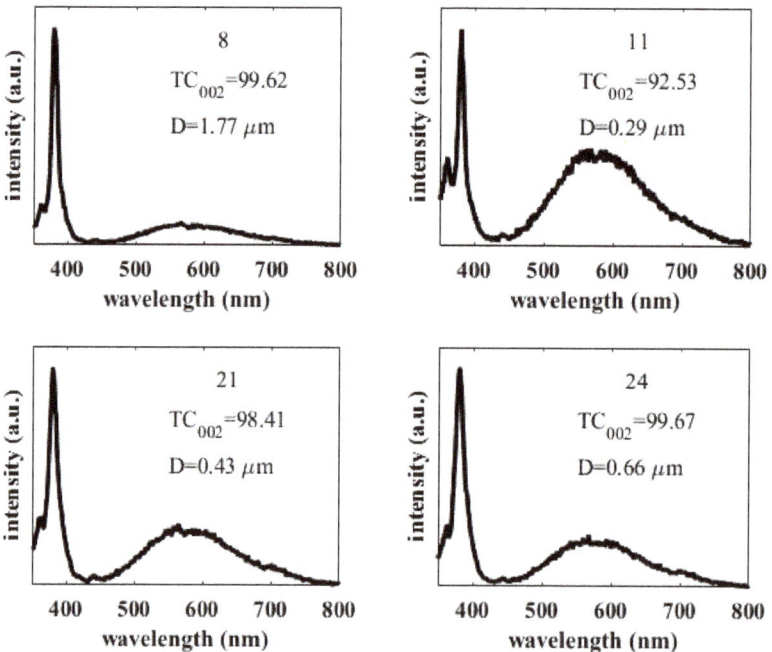

Figure 9. Photoluminescence spectra of the obtained ZnO rod arrays with different parameters. The number denoted above each spectrum represents the experiment number in Table 1.

4. Conclusions

In this study, vertically aligned ZnO rods arrays with large diameters were successfully synthesized by using a chemical bath deposition method. With the assistance of response surface methodology, the oriented growth, aspect ratio, and diameter could be systematically controlled. The changing tendency of the three responses with the involved factors was mapped, and the required parameters to achieve the specific values of response could be efficiently found. With the optimized parameters, the maximum value of texture coefficient and diameter could reach 99.62 and 1.77 µm, respectively. On this condition, the rods were merged together, and a dense ZnO thin film was formed.

Supplementary Materials: The following are available online at http://www.mdpi.com/2227-9717/8/6/655/s1, Figure S1: The SEM image of obtained ZnO rods in the edge of the samples. Figure S2: The enlarged cross-sectional images of the obtained ZnO rod arrays, corresponding to Figure 8 in the manuscript. All authors have read and agreed to the published version of the manuscript

Author Contributions: Conceptualization, Y.C.; data curation, X.S. and Y.C.; formal analysis, X.S. and Z.Z.; investigation, X.S.; methodology, Y.C., Y.Y. and Z.Z.; project administration, Y.C.; writing—original draft, X.S.; writing—review and editing, Y.C., Y.Y., and Z.Z. All authors have read and agreed to the published version of the manuscript.

Funding: The APC was funded by the China Postdoctoral Science Foundation (Grant No. 2019M651945); Research Project of Hubei Provincial Education Department (Q20191802); Doctoral Research Startup Fund in Hubei University of Automotive Technology (K201802); Hubei Provincial Natural Science Foundation of China (2020CFBXXX).

Conflicts of Interest: The authors declare no conflict of interest.

References

1. Xiong, S.; Ma, J.; Volz, S.; Dumitrică, T. Thermally-Active Screw Dislocations in Si Nanowires and Nanotubes. *Small* **2014**, *10*, 1756–1760. [CrossRef] [PubMed]
2. Zhou, Y.; Xiong, S.; Zhang, X.; Volz, S.; Hu, M. Thermal transport crossover from crystalline to partial-crystalline partial-liquid state. *Nat. Commun.* **2018**, *9*, 4712. [CrossRef] [PubMed]
3. Xiong, S.; Selli, D.; Neogi, S.; Donadio, D. Native surface oxide turns alloyed silicon membranes into nanophononic metamaterials with ultralow thermal conductivity. *Phys. Rev. B* **2017**, *95*, 180301. [CrossRef]
4. Liang, Z.; Gao, R.; Lan, J.-L.; Wiranwetchayan, O.; Zhang, Q.; Li, C.; Cao, G. Growth of vertically aligned ZnO nanowalls for inverted polymer solar cells. *Sol. Energy Mater. Sol. Cells* **2013**, *117*, 34–40. [CrossRef]
5. Lai, F.-I.; Yang, J.-F.; Hsu, Y.-C.; Kuo, S.-Y. Omnidirectional light harvesting enhancement of dye-sensitized solar cells decorated with two-dimensional ZnO nanoflowers. *J. Alloy. Compd.* **2020**, *815*, 152287. [CrossRef]
6. Chandrasekhar, P.; Dubey, A.; Qiao, Q. High efficiency perovskite solar cells using nitrogen-doped graphene/ZnO nanorod composite as an electron transport layer. *Sol. Energy* **2020**, *197*, 78–83. [CrossRef]
7. Wang, Z.L.; Zhu, G.; Yang, Y.; Wang, S.; Pan, C. Progress in nanogenerators for portable electronics. *Mater. Today* **2012**, *15*, 532–543. [CrossRef]
8. Jin, C.; Hao, N.; Xu, Z.; Trase, I.; Nie, Y.; Dong, L.; Closson, A.; Chen, Z.; Zhang, J.X. Flexible piezoelectric nanogenerators using metal-doped ZnO-PVDF films. *Sensors Actuators A Phys.* **2020**, *305*, 111912. [CrossRef]
9. Chen, M.-T.; Lu, M.-P.; Wu, Y.-J.; Song, J.; Lee, C.-Y.; Lu, M.-Y.; Chang, Y.-C.; Chou, L.-J.; Wang, Z.L.; Chen, L.J. Near UV LEDs Made with in Situ Doped p-n Homojunction ZnO Nanowire Arrays. *Nano Lett.* **2010**, *10*, 4387–4393. [CrossRef]
10. Deng, G.; Zhang, Y.; Yu, Y.; Han, X.; Wang, Y.; Shi, Z.; Dong, X.; Zhang, B.; Du, G.; Liu, Y. High-Performance Ultraviolet Light-Emitting Diodes Using n-ZnO/p-hBN/p-GaN Contact Heterojunctions. *ACS Appl. Mater. Interfaces* **2020**, *12*, 6788–6792. [CrossRef]
11. Jood, P.; Mehta, R.J.; Zhang, Y.; Peleckis, G.; Wang, X.; Siegel, R.W.; Borca-Tasciuc, T.; Dou, S.; Ramanath, G. Al-Doped Zinc Oxide Nanocomposites with Enhanced Thermoelectric Properties. *Nano Lett.* **2011**, *11*, 4337–4342. [CrossRef] [PubMed]
12. Ong, K.; Singh, D.J.; Wu, P. Analysis of the thermoelectric properties of n-type ZnO. *Phys. Rev. B* **2011**, *83*, 115110. [CrossRef]
13. Xiong, S.; Sääskilahti, K.; Kosevich, Y.A.; Han, H.; Donadio, D.; Volz, S. Blocking Phonon Transport by Structural Resonances in Alloy-Based Nanophononic Metamaterials Leads to Ultralow Thermal Conductivity. *Phys. Rev. Lett.* **2016**, *117*, 025503. [CrossRef] [PubMed]
14. Xiong, S.; Latour, B.; Ni, Y.; Volz, S.; Chalopin, Y. Efficient phonon blocking in SiC antiphase superlattice nanowires. *Phys. Rev. B* **2015**, *91*, 224307. [CrossRef]
15. Sheng, X.; Li, Z.; Cheng, Y. Electronic and Thermoelectric Properties of V2O5, MgV2O5, and CaV2O5. *Coatings* **2020**, *10*, 453. [CrossRef]
16. Menzel, A.; Subannajui, K.; Güder, F.; Moser, D.; Paul, O.; Zacharias, M. Multifunctional ZnO-Nanowire-Based Sensor. *Adv. Funct. Mater.* **2011**, *21*, 4342–4348. [CrossRef]
17. Vinoth, E.; Gopalakrishnan, N. Fabrication of interdigitated electrode (IDE) based ZnO sensors for room temperature ammonia detection. *J. Alloy. Compd.* **2020**, *824*, 153900. [CrossRef]

18. Cheng, Y.; Wang, J.; Jönsson, P.G.; Zhao, Z. Optimization of high-quality vertically aligned ZnO rod arrays by the response surface methodology. *J. Alloy. Compd.* **2015**, *626*, 180–188. [CrossRef]
19. Cheng, Y.; Wang, J.; Jönsson, P.G.; Zhao, Z. Improvement and optimization of the growth quality of upright ZnO rod arrays by the response surface methodology. *Appl. Surf. Sci.* **2015**, *351*, 451–459. [CrossRef]
20. Das, S.; Dutta, K.; Pramanik, A. Morphology control of ZnO with citrate: A time and concentration dependent mechanistic insight. *CrystEngComm* **2013**, *15*, 6349. [CrossRef]
21. Nicholas, N.J.; Franks, G.; Ducker, W.A. Selective Adsorption to Particular Crystal Faces of ZnO. *Langmuir* **2012**, *28*, 7189–7196. [CrossRef] [PubMed]
22. Lifson, M.L.; Levey, C.; Gibson, U.J. Diameter and location control of ZnO nanowires using electrodeposition and sodium citrate. *Appl. Phys. A* **2013**, *113*, 243–247. [CrossRef]
23. Suwanboon, S.; Somraksa, W.; Amornpitoksuk, P.; Randorn, C. Effect of trisodium citrate on the formation and structural, optical and photocatalytic properties of Sr-doped ZnO. *J. Alloy. Compd.* **2020**, *832*, 154963. [CrossRef]
24. Li, Z.; Zhang, L.; He, X.; Bensong, C.; Chen, B. Urchin-like ZnO-nanorod arrays templated growth of ordered hierarchical Ag/ZnO hybrid arrays for surface-enhanced Raman scattering. *Nanotechnology* **2020**, *31*, 165301. [CrossRef] [PubMed]
25. Reinhardt, K.A.; Reidy, R.F. *Handbook for Cleaning for Semiconductor Manufacturing: Fundamentals and Applications*; John Wiley & Sons: Hoboken, NJ, USA, 2011; Volume 67.
26. Xiong, S.; Yang, K.; Kosevich, Y.A.; Chalopin, Y.; D'Agosta, R.; Cortona, P.; Volz, S. Classical to Quantum Transition of Heat Transfer between Two Silica Clusters. *Phys. Rev. Lett.* **2014**, *112*, 114301. [CrossRef] [PubMed]
27. Ouyang, B.; Xiong, S.; Yang, Z.; Jing, Y.; Wang, Y. MoS 2 heterostructure with tunable phase stability: Strain induced interlayer covalent bond formation. *Nanoscale* **2017**, *9*, 8126–8132. [CrossRef] [PubMed]
28. Cheng, Y.; Jönsson, P.G.; Zhao, Z. Controllable fabrication of large-area 2D colloidal crystal masks with large size defect-free domains based on statistical experimental design. *Appl. Surf. Sci.* **2014**, *313*, 144–151. [CrossRef]
29. Liao, Z.-M.; Zhang, H.; Zhou, Y.; Xu, J.; Zhang, J.-M.; Yu, D. Surface effects on photoluminescence of single ZnO nanowires. *Phys. Lett. A* **2008**, *372*, 4505–4509. [CrossRef]

© 2020 by the authors. Licensee MDPI, Basel, Switzerland. This article is an open access article distributed under the terms and conditions of the Creative Commons Attribution (CC BY) license (http://creativecommons.org/licenses/by/4.0/).

Article

Synthesis, Electrical Properties and Na+ Migration Pathways of Na$_2$CuP$_{1.5}$As$_{0.5}$O$_7$

Ohud S. A. ALQarni [1], Riadh Marzouki [1,2,3,*], Youssef Ben Smida [4], Majed M. Alghamdi [1], Maxim Avdeev [5,6], Radhouane Belhadj Tahar [1] and Mohamed Faouzi Zid [3]

1. Chemistry Department, College of Science, King Khalid University, Abha 61413, Saudi Arabia; 438800025@kku.edu.sa (O.S.A.A.); mmalghamdi@kku.edu.sa (M.M.A.); rhbalhaj@kku.edu.sa (R.B.T.)
2. Chemistry Department, Faculty of Sciences of Sfax, University of Sfax, Sfax 3038, Tunisia
3. Laboratoire de Matériaux, Cristallochimie et Thermodynamique Appliquée, Faculté des Sciences de Tunis, Université de Tunis El Manar, El Manar II, Tunis 2092, Tunisia; medfaouzi.zid57@gmail.com
4. Laboratory of valorization of useful materials, National Center of Materials Sciences Research, Technopole Borj Cedria, BP 73, Soliman 8027, Tunisia; youssef.bensmida@fst.utm.tn
5. Australian Centre for Neutron Scattering, Australian Nuclear Science and Technology Organisation, Sydney 2234, Australia; Maxim.Avdeev@ansto.gov.au
6. School of Chemistry, The University of Sydney, Sydney 2006, Australia
* Correspondence: rmarzouki@kku.edu.sa; Tel.: +96-654-284-3436

Received: 10 February 2020; Accepted: 3 March 2020; Published: 6 March 2020

Abstract: A new member of sodium metal diphosphate-diarsenate, Na$_2$CuP$_{1.5}$As$_{0.5}$O$_7$, was synthesized as polycrystalline powder by a solid-state route. X-ray diffraction followed by Rietveld refinement show that the studied material, isostructural with β-Na$_2$CuP$_2$O$_7$, crystallizes in the monoclinic system of the C2/c space group with the unit cell parameters a = 14.798(2) Å; b = 5.729(3) Å; c = 8.075(2) Å; β = 115.00(3)°. The structure of the studied material is formed by Cu$_2$P$_4$O$_{15}$ groups connected via oxygen atoms that results in infinite chains, wavy saw-toothed along the [001] direction, with Na$^+$ ions located in the inter-chain space. Thermal study using DSC analysis shows that the studied material is stable up to the melting point at 688 °C. The electrical investigation, using impedance spectroscopy in the 260–380 °C temperature range, shows that the Na$_2$CuP$_{1.5}$As$_{0.5}$O$_7$ compound is a fast-ion conductor with σ_{350} °C = 2.28 10^{-5} Scm^{-1} and Ea = 0.6 eV. Na$^+$ ions pathways simulation using bond-valence site energy (BVSE) supports the fast three-dimensional mobility of the sodium cations in the inter-chain space.

Keywords: diphosphate-diarsenate; crystal structure; electrical properties; transport pathways simulation

1. Introduction

The research exploration of new inorganic materials with open framework constructed of polyhedra sharing faces, edges and/or corners forming 1D channels, 2D inter-layer spaces or 3D networks where cations are located, is currently an area of intense activity including several disciplines, in particular solid-state chemistry. In particular, alkali metal phosphates were found to have various applications because of their electric, piezoelectric, ferroelectric, magnetic, and catalytic properties [1–4]. Among those, the families of materials with the melilite structure [5], the olivine structure [6] and the natrium super ionic conductor (NaSICON) structure [7], attracted attention for their ionic conduction and exchange of ions [6,7].

More recently, in a series of studies arsenate analogs have been synthesized [8–10]. But, until today phosphate compounds are more studied as cathodes [11,12] compared to arsenate and this is perhaps due to the toxicity of arsenic III (As$_2$O$_3$). However, the oxide of arsenic V (As$_2$O$_5$) is less toxic. In addition, the introduction of arsenic into a structure changes its physical and chemical properties and even toxicity. On the other hand, the comparison of the electrochemical properties of LiCoPO$_4$ and LiCoAsO$_4$ both with olivine structure and close unit cell parameters, shows reversible (de)intercalation from/into material at average voltages of 4.8 and 4.6 V, respectively for LiCoAsO$_4$ [10] and a voltage average of 2.5–5 V for LiCoPO$_4$ [11].

The Na$_2$MP$_2$O$_7$ systems (M = transition metal) [13,14] have a layered structure with the layers [MP$_2$O$_7$]$_n^{2n-}$ and the sodium cations localized in the interlayer space, which favors high ionic conductivity. We recently investigated the effect of the substitution of P by As with a larger ionic radius for ionic conductivity and showed the improvement of ionic conductivity for Na$_2$CoP$_{1.5}$As$_{0.5}$O$_7$ [15], which has an electrical conductivity value of $\sigma_{240\,°C}$ = 7.91×10^{-5} Scm^{-1} and an activation energy Ea = 0.56 eV compared to Na$_2$CoP$_2$O$_7$ ($\sigma_{300\,°C}$ = 2 × 10^{-5} Scm^{-1}; Ea = 0.63 eV) [13].

In our search for new polyanion oxides of sodium and transition metals, the exploration of the Na$_2$O-CuO-P$_2$O$_5$-As$_2$O$_5$ crystallographic systems allowed us to isolate a new member of di-phosphate arsenates, Na$_2$CuP$_{1.5}$As$_{0.5}$O$_7$, in the polycrystalline powder form. In this paper, characterizations and physicochemical studies of the new member of sodium copper diphosphate-diarsenate material, Na$_2$CuP$_{1.5}$As$_{0.5}$O$_7$, and a comparison with other previous works encountered in the literature will be presented.

2. Materials and Methods

A mixture of Cu(NO$_3$)$_2$.2.5H$_2$O, NH$_4$H$_2$PO$_4$ and Na$_2$HAsO$_4$.7H$_2$O in the molar ratio Na:Cu:P:As equal to 2:1:1.5:0.5 was placed in a porcelain crucible and heated to 350 °C for 24 h to eliminate the volatile products H$_2$O, NO$_2$, and NH$_3$. The obtained powder was ground manually using agate mortar and shaped as cylindrical pellets by a uniaxial press. The obtained pellets were heated to 600 °C. After 72 h, the sample was cooled slowly at a rate of 10 °C/h down to room temperature. After grinding finely, a blue polycrystalline powder was obtained.

X-ray powder diffraction (XRD) was used to control and ensure the purity of the obtained powder. The analysis was carried out using XRD-6000 (Shimadzu, Japan) with graphite monochromator (Cukα, λ = 0.154178 nm) and a scan range of 2θ = 10°–70° with step of about 0.02°. The structure was refined using the Rietveld method by the means of the GSAS computer program [16] (EXPGUI, Gaithersburg, Maryland, USA). The crystallographic data of Na$_2$CuP$_2$O$_7$ [17] was used as a starting set. The obtained structural model was confirmed by the Charge Distribution CHARDI model of validation. The CHARDI calculation was done by using the CHARDI2015 computer program (Nespolo, IUCR) [18].

FTIR spectrometer (Agilent Technologies Cary 630 model) was used to allow a direct indexation of the peaks on a spectral range in wave numbers ranging from (1300–400 cm^{-1}).

Differential scanning calorimetry (DSC), with the SDT Q600 model, was used to study the thermal behavior of the obtained and prepared sample. In fact, the device contains two crucibles, one as a reference and the other contains the sample to be analyzed. These two crucibles are heated to 750 °C at a rate of 10 °C/min. The thermal analysis was carried out under a nitrogen atmosphere to avoid the reaction of the sample with the oxygen in the air.

Energy-dispersive X-ray spectroscopy (EDX) and scanning electron microscopy (SEM, Thermo Fisher Scientific model), were used to identify the present elements and the microstructure of the studied material, respectively.

The electrical measurements were preceded by pretreatment of the sample in order to densify the measured sample by reducing the mean particle size of the synthesized powder. Mechanical grinding for 100 min was carried out using the FRISCH planetary micromill pulverisette 7. The polycrystalline sample was shaped as a cylindrical pellet using a uniaxial press. The pellet was sintered in air at an optimal temperature of 610 °C for 2 h with 5 °C/min heating and cooling rates. The geometric factor of the dense ceramic is g = e/S = 0.793 cm^{-1} where e and S are the thickness and face area of pellet, respectively. Gold metal electrodes ~36 nm thick were deposited using a SC7620 mini sputter coater.

The sample was then placed between two platinum electrodes that were connected by platinum cables to the frequency response analyzer (HP 4192A) which was controlled by a microcomputer.

Impedance spectroscopic measurements were performed via the Hewlett-Packard 4192-A automatic bridge supervised by HP workstation. Impedance spectra were recorded with 0.5 V AC-signal in the 5 Hz–13 MHz frequency range.

The bond-valence site energy (BVSE) model [19,20] was used to simulate the alkali migration in the 3D anionic framework. The BVSE model is the latest extension of the bond-valence sum (BVS) model developed by Pauling [21] to describe the formation of inorganic materials. The BVS model was improved by Brown & Altermatt [22] followed by Adams [23], resulting in the expression:

$$s_{A-X} = \exp\left(\frac{R_0 - R_{A-X}}{b}\right) \quad (1)$$

where s_{A-X} is individual bond-valence, R_{A-X} is the distance between counter-ions A and X, R_0 and b are fitted empirical constants, and R_0 is the length of a bond of unit valence.

The BVSE model was extensively used in the cation motion simulation in the anionic framework by following the valence unit as a function of migration distance [24]. The valence unit was also recently related to potential energy scale and electrostatic interactions [19,20]. The BVSE method was used with success to simulate the transport pathways of monovalent cations (Na^+; K^+ and Ag^+) in numerous materials including $Na_2CoP_{1.5}As_{0.5}O_7$ [15], $Na_{1.14}K_{0.86}CoP_2O_7$ [25] and $Ag_{3.68}Co_2(P_2O_7)_2$ [26]. The BVSE calculations were performed using the SoftBV [27] code and the visualization of isosurfaces was carried out using VESTA3 software (version 3, Koichi Momma and Fujio Izumi, 2018).

3. Results and Discussion

3.1. X-ray Powder Diffraction

The crystallographic study was started by a simple comparison between the XRD pattern of the prepared materials in the Na_2O-CuO-P_2O_5-As_2O_5 system and those of the previous studies of diphosphate $Na_2MP_2O_7$ [5,7,14,28,29] and $Na_2CoP_{1.5}As_{0.5}O_7$ [15]. In this case, only the $Na_2CuP_{1.5}As_{0.5}O_7$ diffractogram showed a similarity with that of the β-$Na_2CuP_2O_7$ diphosphate [17]. It crystallizes in the monoclinic system of the C2/c space group. This result prompted us to make a precise refinement using the Rietveld method which was implemented into the GSAS computer software [16]. The final agreement factors are Rp = 5.4% and Rwp = 6.9%. No additional peaks were detected. The final Rietveld plot is presented in Figure 1. The unit cell parameters obtained from the Rietveld refinement are a = 14.798(2) Å; b = 5.729(3) Å; c = 8.075(2) Å; β = 115.00(3)° (Table 1). The details of the crystallographic data, data collection and final agreement factors are given in Table 2. The atomic coordinates and isotropic displacement parameters are listed in Table 3. The main bond distances are given in Table 4. The charge distribution analysis and the bond-valence computation are summarized in Table 5.

Table 1. Unit cell parameters of the β-$Na_2CuP_2O_7$ and $Na_2CuP_{1.5}As_{0.5}O_7$ materials.

Parameter	$Na_2CuP_{1.5}As_{0.5}O_7$ (Current Work)	β-$Na_2CuP_2O_7$ [30]
a (Å)	14.798(2)	14.728(3)
b (Å)	5.729(3)	5.698(1)
c (Å)	8.075(2)	8.067(1)
β (°)	115.00(3)	115.15(1)
V (Å^3)	620.43(3)	612.80(2)

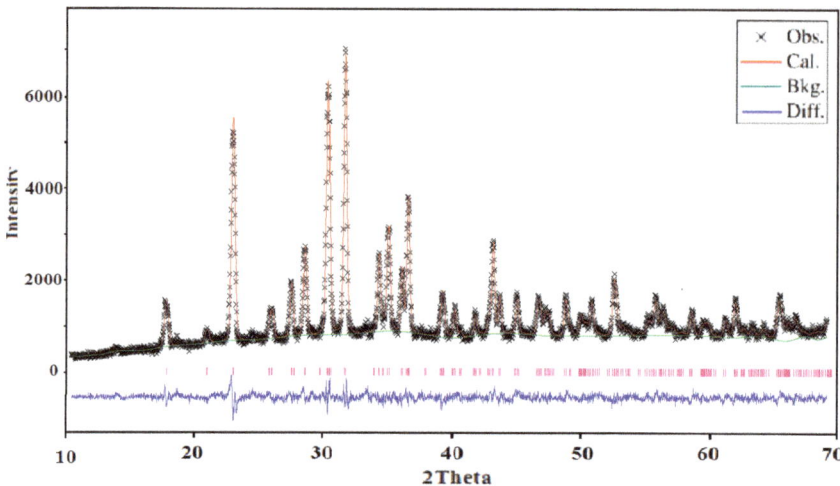

Figure 1. Results of the Rietveld refinement of the powder of $Na_2CuP_{1.5}As_{0.5}O_7$ based on XRD data.

Table 2. Structure refinement results of the $Na_2CuP_{1.5}As_{0.5}O_7$ compound.

Crystallographic Data	
Empirical Formula	$Na_2CuP_{1.5}As_{0.5}O_7$
Formula Weight; ρ_{cal}	305.44 g mol^{-1}; 3.220 g cm^{-1}
Crystalline System; Space Group	Monoclinic, C2/c
Unit Cell Dimensions	a = 14.8688 (8), b = 5.7591 (3), c = 13.5957 (7) β = 147.2406 (12)
Volume; Z	V = 629.97 (6) Å^3; 4
Data Collection	
Diffractometer	Bruker D8 ADVANCE
Wavelength	$\lambda_{Cu\,K\alpha}$ = 1.54056 Å
Temperature	298 (2) K
Angle Range	4.91°–69.91°
Step Scan Increment (°2θ)	0.02°
Counting Time	2 s
Refinement	
Angle Range	4.91°–69.91°
R_p	0.054
R_{wp}	0.069
R_{exp}	0.043
$R(F^2)$	0.05117
Goodness of Fit $\chi 2$	2.592
No. of Data Points	3251
No. of Restraints	18
Profile Function	Pseudo-Voigt
Background	Chebyshev function with 20 terms

Table 3. Fractional atomic coordinates and equivalent isotropic displacement parameters (Å^2).

	x	y	z	U_{iso}	Occ. (<1)
Cu1	$\frac{1}{4}$	$\frac{1}{4}$	$\frac{1}{2}$	0.0123 (12)	
P1/As1	0.5112 (2)	0.5884 (5)	0.6563 (3)	0.0065 (14)	0.75/0.25
Na1	0.3211 (5)	0.8919 (9)	0.2991 (5)	0.004 (2)	
O1	0.6590 (6)	0.4179 (9)	0.7999 (6)	0.006 (2)	
O2	0.5389 (4)	0.7622 (10)	0.6011 (6)	0.006 (2)	
O3	0.8432 (6)	0.0371 (12)	0.9916 (5)	0.041 (5)	
O4	$\frac{1}{2}$	0.7267 (9)	$\frac{3}{4}$	0.018 (5)	

Table 4. Main bond distances (Å) in the coordination polyhedra for $Na_2CuP_{1.5}As_{0.5}O_7$.

CuO_4		$(P1/As1)O_4$	
Cu1—O1^{iii}	1.9938 (3)	(P1/As1)—O1	1.5457 (3)
Cu1—O1^{x}	1.9938 (3)	(P1/As1)—O2	1.5015 (3)
Cu1—O3^{iii}	1.9247 (3)	(P1/As1)—O3^{x}	1.5387 (3)
Cu1—O3^{x}	1.9247 (3)	(P1/As1)—O4	1.6248 (3)
$Na1O_6$			
Na1—O1^{i}	2.3985 (4)	Na1—O2^{iv}	2.3220 (4)
Na1—O1^{vi}	2.6566 (4)	Na1—O2^{vi}	2.5780 (4)
Na1—O2	2.3541 (4)	Na1—O3^{i}	2.3485 (4)

Symmetry codes: (i) −x + 1, −y + 1, −z + 1; (ii) x−1/2, −y + 1/2, z + 1/2; (iii) −x + 1, y, −z + 3/2; (iv) −x + 1, −y + 2, −z + 1; (v) x−1/2, −y−1/2, z + 1/2; (vi) x−3/2, −y + 1/2, z−1/2; (vii) −x + 1/2, y−1/2, −z + 1/2; (viii) −x + 1/2, y + 1/2, −z + 1/2; (ix) x, −y + 1, z−1/2; (x) x−3/2, −y−1/2, z−1.

Table 5. Charge distribution analysis of cation polyhedra in $Na_2CuP_{1.5}As_{0.5}O_7$.

Cation	q(i).sof(i)	Q(i)	CN(i)	ECoN(i)	$d_{ar}(i)$	$d_{med}(i)$
Cu1	2.000	1.96	4	3.96	1.955	1.959
M1	5.000	5.03	4	3.88	1.544	1.552
Na1	1.000	0.98	6	5.47	2.400	2.443

q(i), formal oxidation number; Q(i), computed charge; sof(i), site occupation factor; $d_{ar}(i)$, arithmetic average distance; $d_{med}(i)$, weighted average distance; CN, coordination number; ECoN(i), effective coordination number. M1 = (0.75P + 0.25As=).

By comparing the unit cell parameters of the studied material with those of β-$Na_2CuP_2O_7$, the P/As substitution effect increases the volume of the unit cell (Table 1), which is explained by the distance of As—O bonds being greater than that of P—O.

3.2. Infrared Spectroscopy

The IR absorption spectrum of the studied $Na_2CuP_{1.5}As_{0.5}O_7$ material is shown in Figure 2. The spectrum shows the presence of the series of distinct bands attributed to asymmetric and symmetrical valence vibrations of the P-O-P and As-O-As bridges. These bands are characteristic of the pyrophosphate $(P_2O_7)^{4-}$ and diarsenate $(As_2O_7)^{4-}$ groups (Table 6) [30] and similar to those of the $Li_2CuP_2O_7$ spectrum [31].

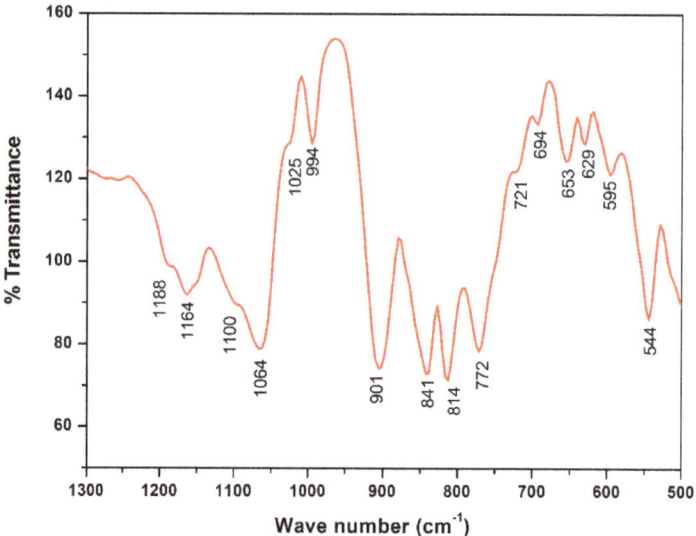

Figure 2. FT-IR spectrum (1300–500 cm^{-1}) of Na$_2$CuP$_{1.5}$As$_{0.5}$O$_7$.

Table 6. Proposed assignment of the vibration bands of Na$_2$CuP$_{1.5}$As$_{0.5}$O$_7$.

Attribution	Wave Number (cm^{-1})
ν_{as} (PO$_3$)	1188
	1164
ν_{as} (AsO$_3$)	1100
ν_s (PO$_3$)	1064
	1025
ν_s (AsO$_3$)	994
ν_{as} (POP) Stretching Vibrations	901
ν_{as} (AsOAs) Stretching Vibrations	841
	814
ν_s (POP) Stretching Vibrations	772
	721
ν_s (AsOAs) Stretching Vibrations	694
	653
δ_{as} (PO$_3$) Deformation Modes	629
	595
δ_s (PO$_3$) Deformation Modes	544

3.3. DSC Thermal Analysis

In order to determine the thermal stability of the studied compound, the DSC analysis was used in the range from room temperature to 750 °C. The analysis result is illustrated in Figure 3. An endothermic peak was observed at 688 °C. This peak corresponds to the melting point of our compound. While, an exothermal peak is shown at 743 °C, after the fusion, probably corresponds to the oxidation of fractions of Cu^{2+} to Cu^{3+} in the obtained liquid phase. Overall, the thermal analysis via DSC shows that $Na_2CuP_{1.5}As_{0.5}O_7$ material is stable up to a temperature of 674 °C. The sharpness of the endothermic peak in the DSC analysis suggests good crystallinity of our synthesized powder.

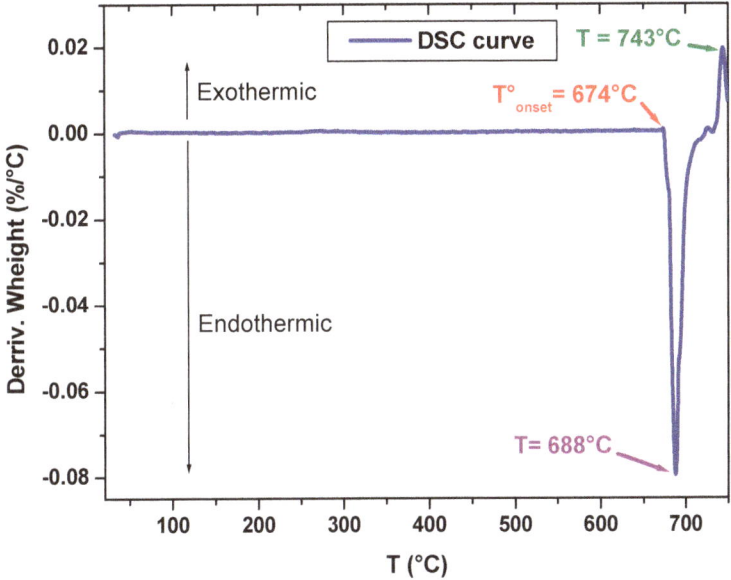

Figure 3. DSC curve of the $Na_2CuP_{1.5}As_{0.5}O_7$ compound.

Here we can also compare the thermal stability of $Na_2CuP_{1.5}As_{0.5}O_7$ to that of the recently studied Co analog $Na_2CoP_{1.5}As_{0.5}O_7$. The Cu material is stable from room temperature to the melting temperature, which is around 688 °C. In contrast, the $Na_2CoP_{1.5}As_{0.5}O_7$ material undergoes a phase transition at a temperature of 675 °C before melting at ~700 °C. This shows that the $Na_2CuP_{1.5}As_{0.5}O_7$ material is more stable than the $Na_2CoP_{1.5}As_{0.5}O_7$ material [15].

3.4. SEM Microstructure and EDX Analysis of $Na_2CuP_{1.5}As_{0.5}O_7$

Energy-dispersive X-ray spectroscopy (EDX) and scanning electron microscopy (SEM) analysis were used to confirm the chemical composition and examine polycrystalline morphology, respectively (Figure 4). The EDX analysis of the polycrystalline powder revealed the presence of the expected elements, i.e., sodium, copper, phosphorus, arsenic, and oxygen. The micrograph on SEM of the sample shows agglomeration of uniform parallelepiped crystallites. The mapping elemental analysis of $Na_2CuP_{1.5}As_{0.5}O_7$ confirmed the uniform distribution of the constituent elements (Figure 5).

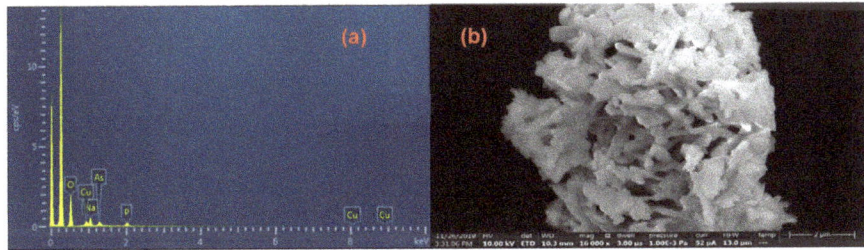

Figure 4. (**a**) EDX analysis and (**b**) SEM micrograph of the $Na_2CuP_{1.5}As_{0.5}O_7$ sample.

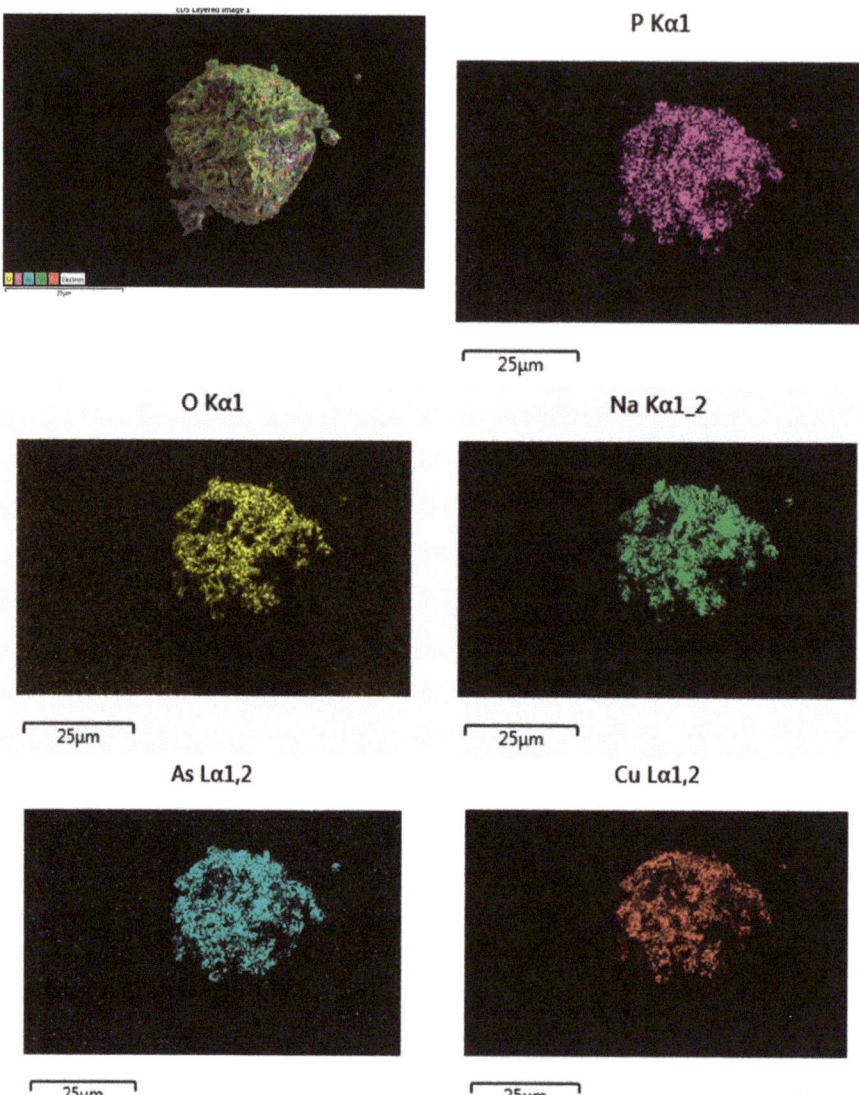

Figure 5. The mapping elemental analysis of the $Na_2CuP_{1.5}As_{0.5}O_7$ sample.

3.5. Crystal Structure Description

The structural unit of $Na_2CuP_{1.5}As_{0.5}O_7$ is presented in Figure 6. It contains two P_2O_7 units connected by a vertex with two CuO_4 of square planar geometry. The charge neutrality of the structural unit is ensured by four sodium ions (Na^+).

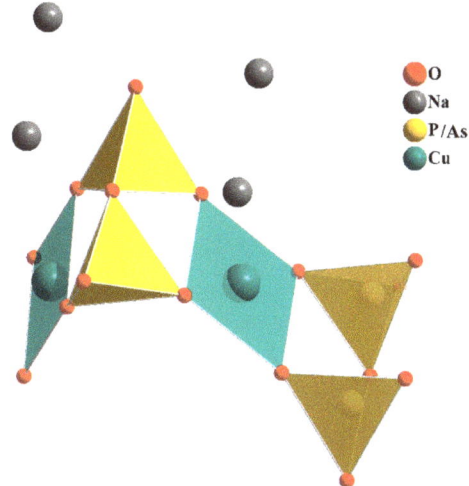

Figure 6. The structural unit of $Na_2CuP_{1.5}As_{0.5}O_7$.

The $Cu_2P_4O_{15}$ groups of the structural unit are bound by oxygen peaks to result in infinite chains and are wavy saw-toothed along the [001] direction (Figure 7). The Na$^+$ ions are located in the inter-chain space.

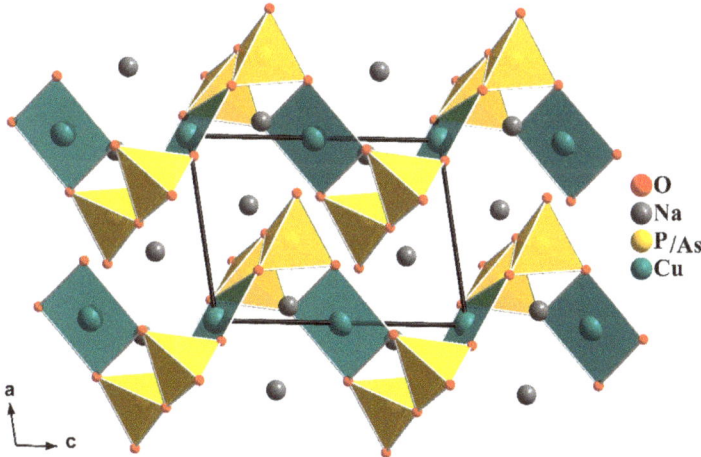

Figure 7. View of the structure of $Na_2CuP_{1.5}As_{0.5}O_7$ in the *ac* plane showing the arrangement of the chains.

Other projections of the structure of $Na_2CuP_{1.5}As_{0.5}O_7$ according to the [100] and [001] directions are shown in Figure 8.

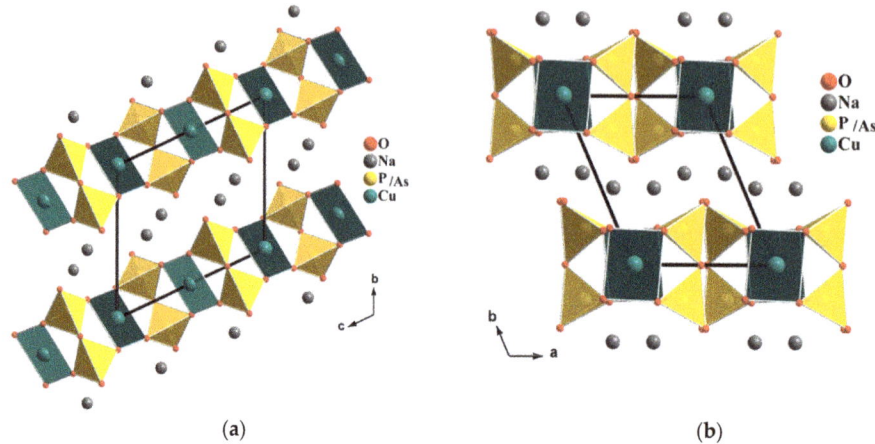

Figure 8. Projections of the Na$_2$CuP$_{1.5}$As$_{0.5}$O$_7$ structure along the (**a**) [100] and (**b**) [001] directions.

The structure of our material differs from that of the allotropic form α-Na$_2$CuP$_2$O$_7$ [17], which has a two-dimensional anionic framework formed by the connection of vertices of PO$_4$ tetrahedra, and CuO$_5$ polyhedra.

Compared to the sodium cobalt diphosphate-diarsenate Na$_2$CoP$_{1.5}$As$_{0.5}$O$_7$ investigated recently by Marzouki et al. [15], we notice that despite a similar composition, Na$_2$CuP$_{1.5}$As$_{0.5}$O$_7$ crystallizes in a different structure type. Indeed, the cobalt material crystallizes in the tetragonal system of the P4$_2$/mnm space group with the unit cell parameters a = 7.764(3) Å, c = 10.385(3) Å. In contrast, the studied material Na$_2$CuP$_{1.5}$As$_{0.5}$O$_7$, crystallizes in the monoclinic system of the C2/c space group with the unit cell parameters a = 14.798(2) Å; b = 5.729(3) Å; c = 8.075(2) Å; β = 115.00(3)°. The difference is undoubtedly determined by the preference of the Jahn–Teller active d^9 Cu^{2+} to adopt square coordination (Figure 6).

3.6. Electrical Properties: Effect of P/As Doping

The prepared pellet of the Na$_2$CuP$_{1.5}$As$_{0.5}$O$_7$ compound was sintered at 550 °C for 2 h with a 5 °C/min heating and cooling rate. The relative density of the obtained pellet is D = 88%. The thickness and surface of the pellet are e = 0.36 cm and S = 0.454 cm^2, respectively. The electrical measurements of the obtained sample were carried out using complex impedance spectroscopy in the temperature range of 260–380 °C. The recorded spectra are shown in Figure 9.

The best fits of impedance spectra were obtained when using a conventional electrical circuit R$_g$//CPE$_g$-R$_{gb}$//CPE$_{gb}$, where CPE are constant phase elements (Figure 9a) and subscripts g and gb indicate bulk grain and grain boundary contribution, respectively:

$$ZCPE = \frac{1}{A(jO))^p} \quad (2)$$

The true capacitance was calculated from the pseudo-capacitance according to the following relationships:

$$O_0 = (RA)^{-1/p} = (RC)^{-1} \quad (3)$$

(where ω_0 is the relaxation frequency, A is the pseudo-capacitance obtained from the CPE, and C is the true capacitance.

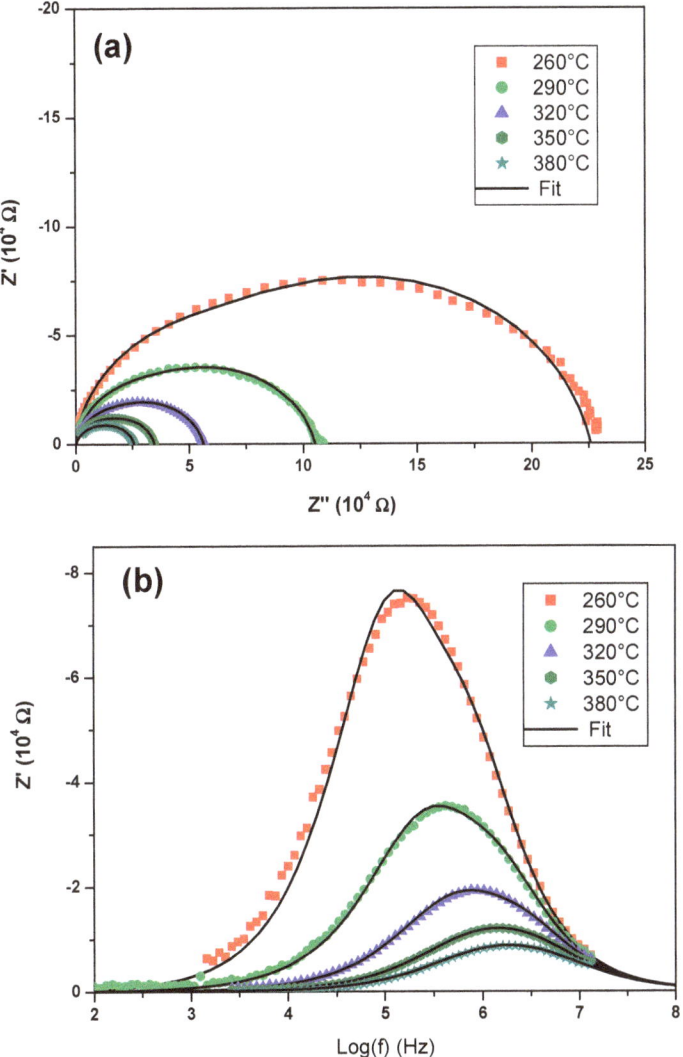

Figure 9. Impedance spectra of $Na_2CuP_{1.5}As_{0.5}O_7$ recorded in a temperature range of 240–360 °C in air. The refined calculated model is shown in (**a**) the Nyquist and (**b**) Bode planes.

The electrical parameter values calculated at different temperatures are shown in Table 4. The values of the capacities C_gk and $C_{gb}k$ are approximately 10^{-11} and 10^{-10} Fcm^{-1} for the bulk and grain boundaries, respectively [15]. In fact, with a relative density of D = 88%, the conductivity of the prepared sample (Table 7) increases from 0.35 10^{-5} Scm^{-1} at 260 °C to 3.13 10^{-5} Scm^{-1} at 380 °C. On the other hand, the 12% porosity of our sample prompted us to estimate the conductivity values of the fully dense sample of $Na_2CuP_{1.5}As_{0.5}O_7$ using the empirical formula proposed by Langlois and Coeuret [32]:

$$\sigma = \frac{(1-P)}{4}\sigma_d \qquad (4)$$

where σ and σ_d are the electrical conductivity of porous and dense samples, respectively. P is the porosity of the sample.

Table 7. Electrical parameters values of equivalent circuits of $Na_2CuP_{1.5}As_{0.5}O_7$ determined by impedance spectroscopy at 260–380 °C.

T (°C)	T (K)	R_g (10^4 Ω)	A_g (10^{-10} F s^{p-1})	C_gk (10^{-11} Fcm^{-1})	R_{gb} (10^4 Ω)	A_{gb} (10^{-10} F s^{p-1})	C_{gb}k (10^{-10} Fcm^{-1})	Rt (10^4 Ω)	ρt (10^4 Ω cm)	σ (10^{-5} S cm^{-1})	$σ_d$ (10^{-5} S cm^{-1})
260	533	5.54	2.1	2.0	17.04	1.7	1.7	22.58	28.58	0.35	1.59
290	563	2.70	3.4	3.3	7.82	1.8	1.70	10.52	13.32	0.75	3.41
320	593	1.43	3.4	3.3	4.18	1.5	1.5	5.61	7.10	1.41	6.41
350	623	1.21	3.4	3.3	2.26	1.4	1.3	3.47	4.39	2.28	10.36
380	653	1.01	8.4	8.0	1.51	1.2	1.2	2.52	3.19	3.13	14.23

This correction has been used in previous works such as $Na_2CoP_{1.5}As_{0.5}O_7$ [15]. Taking into account the porosity factor P = 0.12, the conductivity value of dense material will be σ_d = (4σ/0.88). The conductivity values of dense sample calculated at different temperatures are summarized in Table 7. In this case, the experimental conductivity of 3.5 10^{-6} Scm^{-1} corresponds to the corrected value of 1.59 10^{-5} Scm^{-1} at 260 °C.

The curve Ln ($\sigma \times$ T) = f (1000/T) is linear (Figure 10), satisfying the Arrhenius law LnσT = Lnσ_0 – Ea/kT (k = Boltzmann constant). The activation energy calculated from the slope of this curve is Ea = 0.60 eV.

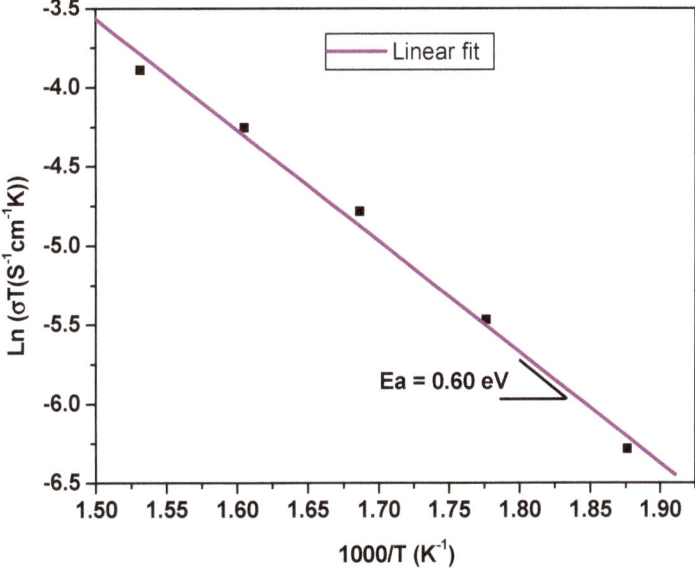

Figure 10. Arrhenius plot of the conductivity of the $Na_2CuP_{1.5}As_{0.5}O_7$ sample.

The electrical investigation of the studied material shows that the activation energy, which is unaffected by porosity and thus easier to use for comparison, decreases for $Na_2CuP_{1.5}As_{0.5}O_7$ compared to that of $Na_2CuP_2O_7$ [33], i.e., 0.60 eV and 0.89 eV, respectively. Consequently, the effect of P/As substitution increases the electrical conductivity of the parent material $Na_2CuP_2O_7$ at lower temperatures [33]. Overall, a comparison of the conductivity values of the studied material $Na_2CuP_{1.5}As_{0.5}O_7$ (at T = 350 °C, $\sigma_{D\,=\,88\%}$ = 2.28 × 10^{-5} Scm^{-1}; σ_d = 2.28 × 10^{-4} Scm^{-1} and Ea = 0.60 eV) with those found in the literature shows that our material can be classified among the fast ionic conductors as shown in Table 8.

Table 8. Activation energies of ionic conductivity for some sodium-ion materials.

Material	Activation Energy (eV)	Temperature Range (°C)	Ref.
$Na_2CuP_{1.5}As_{0.5}O_7$	0.60	260–380	current work
$Na_2CoP_{1.5}As_{0.5}O_7$	0.56	240–360	[15]
$Na_{1.14}K_{0.86}CoP_2O_7$	1.34	360–480	[24]
$Na_{2.84}Ag_{1.16}Co_2(P_2O_7)_2$	1.36	510–630	[34]
$NaCo_2As_3O_{10}$	0.48	160–410	[35]
$Na_4Co_{5.63}Al_{0.91}(AsO_4)_6$	0.56	400–550	[36]
$Na_2Co_2(MoO_4)_3$	1.20	180–513	[37]

3.7. BVSE Simulation: Na^+ Migration Pathways in $Na_2CuP_{1.5}As_{0.5}O_7$

The BVSE calculation revealed in addition to the equilibrium site Na1, the presence of two interstitials sites (i1 to i2) and ten saddle points (s1 to s10) (Table 9). Thus, there are ten local pathways as shown in Table 10. Figure 11 shows the position of equilibrium sites and interstitial sites in the unit cell.

Table 9. Bond-valence sites energies and positions of equilibrium site (Na1), interstitial sites (i1 and i2) and saddle points (s1 to s10).

Site	x	y	z	Energy (eV)
Na1	0.185	0.611	0.701	0.000
i1	0.524	0.875	0.424	0.288
i2	0.262	0.194	0.986	0.715
s1	0.560	0.125	0.632	0.354
s2	0.226	0.083	0.215	0.466
s3	0.006	0.389	0.250	0.550
s4	0.125	0.528	0.069	0.602
s5	0.250	0.250	0.000	0.749
s6	0.173	0.431	0.958	0.870
s7	0.714	0.208	0.583	0.960
s8	0.821	0.708	0.375	1.260
s9	0.190	0.333	0.653	1.289
s10	0.167	0.361	0.833	1.643

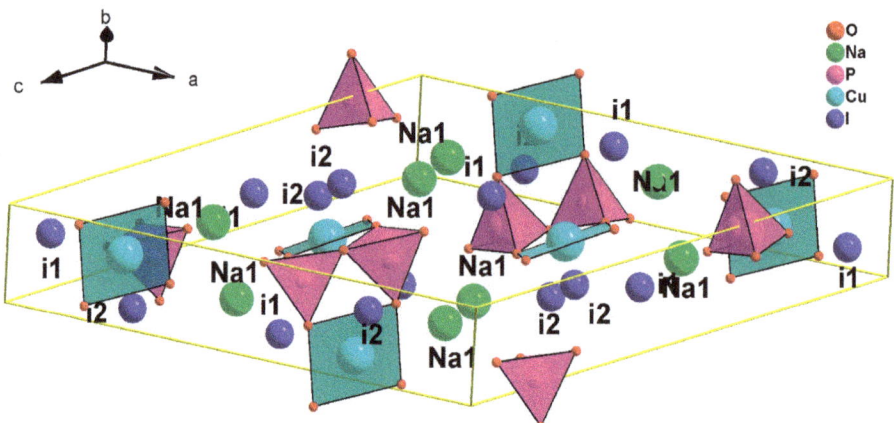

Figure 11. Unit cell of the title compound showing the position of the equilibrium site Na1 and the interstitial sites (i1 and i2).

Table 10. Local transport pathways in the anionic framework, energetic barrier (eV) and hop distance (Å).

Local Path	Site 1	Saddle	Site 2	Barrier (eV)	Hop Distance (Å)
1	i1	s1	Na1	0.354	1.904
2	Na1	s2	Na1	0.466	3.119
3	i1	s3	i1	0.261	4.145
4	i1	s4	Na1	0.602	2.593
5	i2	s5	i2	0.033	0.955
6	i2	s6	i1	0.582	2.922

Table 10. Cont.

Local Path	Site 1	Saddle	Site 2	Barrier (eV)	Hop Distance (Å)
7	i2	s7	Na1	0.960	3.282
8	i1	s8	Na1	1.260	3.168
9	Na1	s9	i1	1.289	3.168
10	i2	s10	i1	1.355	3.956

Figure 11 shows that the migration along the b direction does not involve any interstitial sites, and the diffusion occurs from the equilibrium site Na1 to its symmetry image, with a jump distance of approximately 3.119 Å and with an activation energy of approximately 0.466 eV (Figure 12a).

Along the c direction, Figure 11 shows that the sodium moves from the Na1 position to the interstitial sites i2 then to i1 to reach the equivalent Na1 site (Figure 12b) with an activation energy along this direction of approximately 0.96 eV (Figure 12b).

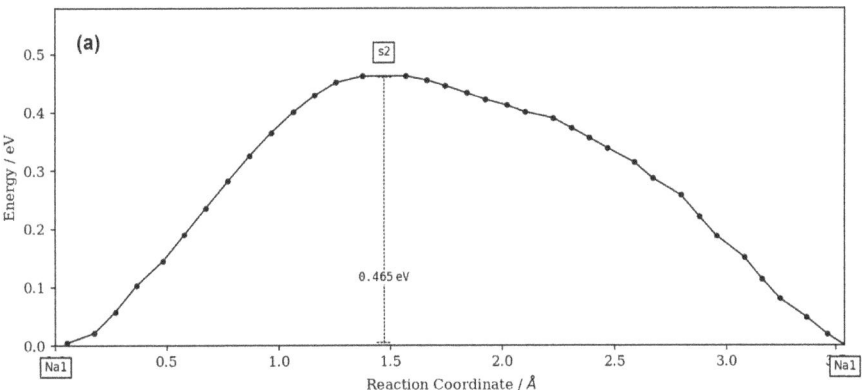

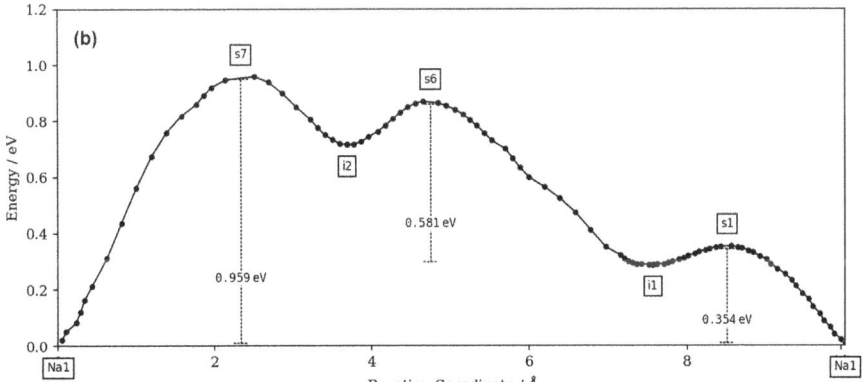

Figure 12. Cont.

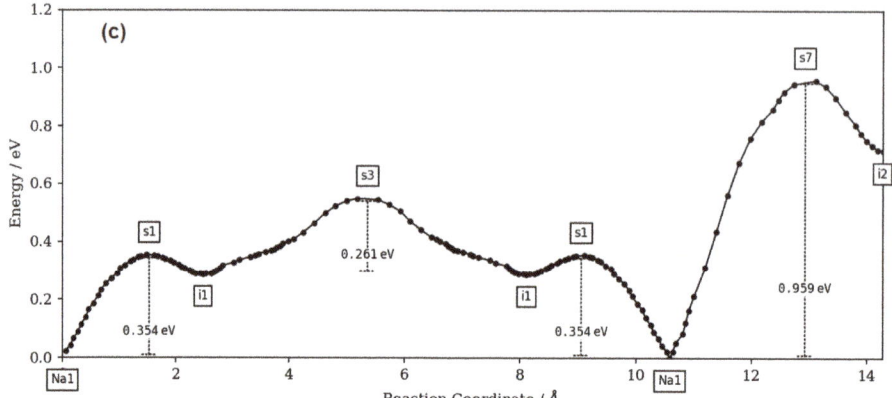

Figure 12. Variation of energy as a function of the reaction coordinate along the (**a**) *b*-direction, (**b**) *c*-direction and (**c**) *a*-direction.

Along the *a* direction, the sodium atoms pass through the following sites: Na1-i1-i1-Na1-i2. The activation energy along this direction is approximately 0.96 eV (Figure 12c). Thus, the activation energy of the title compound for 1D and 3D ionic conductivity is approximately 0.466 eV and 0.96 eV, respectively. Figure 13 shows the isosufaces of conduction pathways.

Figure 13. Isosurfaces of conduction showing the polyhedral of coordination and the 3D ionic conductivity pathways transport of sodium in $Na_2CuP_{1.5}As_{0.5}O_7$.

Consequently, based on the BVSE calculations, the fast ionic conductivity observed for the material can be explained by the three-dimensional mobility of Na^+ ions in the inter-ribbon space, likely with more favorable diffusion along the b-axis.

4. Conclusions

A new quaternary oxide $Na_2CuP_{1.5}As_{0.5}O_7$ was identified in the Na_2O-CuO-P_2O_5-As_2O_5 system. It crystallizes in the monoclinic C2/c space group and is isostructural to β-$Na_2CuP_2O_7$. The partial substitution of P appears to be beneficial for ionic conductivity as the material exhibits lower activation energy of 0.6 eV vs. 0.89 eV for the parent β-$Na_2CuP_2O_7$. According to the impedance spectroscopy performed on the 88% dense pellet, the bulk ionic conductivity reaches the value of $2.28 \cdot 10^{-5}$ Scm^{-1}, which allows $Na_2CuP_{1.5}As_{0.5}O_7$ to classify as a fast ion conductor. The bond-valence site energy calculations suggest that the Na$^+$ diffusion is three-dimensional with some preference for transport along the *b* axis.

Author Contributions: Investigation, O.S.A.A., M.M.A. and M.A.; methodology, R.M.; software, Y.B.S.; supervision, R.M. and R.B.T.; validation, M.F.Z.; visualization, M.M.A.; writing—original draft, O.S.A.A. and R.M.; writing—review and editing, Y.B.S., M.A., R.B.T. and M.F.Z. All authors have read and agreed to the published version of the manuscript.

Funding: This research was funded by the Deanship of Scientific Research at King Khalid University, grant number R.G.P.1/150/40.

Acknowledgments: The authors extend their appreciation to the Deanship of Scientific Research at King Khalid University for funding this work through the research group program under grant number R.G.P.1/150/40.

Conflicts of Interest: The authors declare no conflict of interest.

References

1. Nagpure, M.; Shinde, K.N.; Kumar, V.; Ntwaeaborwa, O.M.; Dhoble, S.J.; Swart, H.C. Combustion synthesis and luminescence investigation of Na$_3$Al$_2$(PO$_4$)$_3$: RE (RE = Ce^{3+}, Eu^{3+} and Mn^{2+}) phosphor. *J. Alloys Compd.* **2010**, *492*, 384–388. [CrossRef]
2. Chen, J.G.; Ang, L.; Wang, C.; Wei, Y. Sol-Gel Preparation and Electrochemical Properties of Na$_3$V$_2$(PO$_4$)$_2$F$_3$ Composite Cathode Material for Lithium Ion Batteries. *J. Alloys Compd.* **2009**, *478*, 604–607.
3. Kanazawa, T. *Inorganic Phosphate Materials, Materials Science Monographs 52*; Elsevier: Amsterdam, The Netherlands, 1989.
4. Hong, S.Y.P. Crystal structures and crystal chemistry in the system Na$_{1+x}$Zr$_2$Si$_x$P$_{3-x}$O$_{12}$. *Mater. Res. Bull.* **1976**, *11*, 173. [CrossRef]
5. Erragh, F.; Boukhari, A.; Elouadi, B.; Holt, E.M. Crystal Structures of Two Allotropic Forms of Na2CoP2O7. *J. Cryst. Spectrosc.* **1991**, *21*, 321–326. [CrossRef]
6. Padhi, A.K.; Nanjundaswamy, K.S.; Goodenough, J.B. Phospho-olivines as Positive-Electrode Materials for Rechargeable Lithium Batteries. *J. Electrochem. Soc.* **1997**, *144*, 1188–1194. [CrossRef]
7. Goodenough, J.B.; Hong, H.Y.-P.; Kafalas, J.A. Fast Na$^+$-ion transport in skeleton structures. *Mater. Res. Bull.* **1976**, *11*, 203–220. [CrossRef]
8. Kobashi, D.; Kohara, S.; Yamakawa, J.; Kawahara, A. Un Monophosphate Synthétique de Sodium et de Cobalt: Na$_4$Co$_7$(PO$_4$)$_6$. *Acta Cryst.* **1998**, *C54*, 7–9. [CrossRef]
9. Marzouki, R.; Frigui, W.; Guesmi, A.; Zid, M.F.; Driss, A. β-Xenophyllite-type Na$_4$Li$_{0.62}$Co$_{5.67}$Al$_{0.71}$(AsO$_4$)$_6$. *Acta Cryst.* **2013**, *E69*, i65–i66. [CrossRef]
10. Arroyo-de Dompablo, M.E.; Amadora, U.; Alvareza, M.; Gallardo, J.M.; García-Alvarado, F. Novel olivine and spinel LiMAsO$_4$ (M = 3d-metal) as positive electrode materials in lithium cells. *Solid State Ion.* **2006**, *177*, 2625–2628. [CrossRef]
11. Prabu, S.M.; Selvasekarapandian, M.V.; Reddy, B.V.R. Chowdari. Impedance studies on the 5-V cathode material, LiCoPO4. *J. Solid State Electrochem.* **2012**, *16*, 1833–1839. [CrossRef]
12. Reddy, M.V.; Rao, G.V.S.; Chowdari, B.V.R. Metal Oxides and Oxysalts as Anode Materials for Li Ion Batteries. *Chem. Rev.* **2013**, *113*, 5364–5457. [CrossRef] [PubMed]
13. Sanz, F.; Parada, C.; Rojo, J.M.; Ruiz-Valero, C.; Saez-Puche, R. Studies on Tetragonal Na$_2$CoP$_2$O$_7$, a Novel Ionic Conductor. *J. Solid State Chem.* **1999**, *145*, 604–611. [CrossRef]
14. Chouaib, S.; Ben Rhaiem, A.; Guidara, K. Dielectric relaxation and ionic conductivity studies of Na$_2$ZnP$_2$O$_7$. *Bull. Mater. Sci.* **2011**, *34*, 915–920. [CrossRef]
15. Marzouki, R.; Ben Smida, Y.; Sonni, M.; Avdeev, M.; Zid, M.F. Synthesis, structure, electrical properties and Na+ migration pathways of Na$_2$CoP$_{1.5}$As$_{0.5}$O$_7$. *J. Solid State Chem.* **2019**, in press. [CrossRef]
16. Toby, B.J. EXPGUI, a graphical user interface for GSAS. *J. Appl. Crystallogr.* **2001**, *34*, 210–213. [CrossRef]
17. Erragh, F.; Boukhari, A.; Abraham, F.; Elouadi, B. The crystal structure of α- and β-Na$_2$CuP$_2$O$_7$. *J. Solid State Chem.* **1995**, *120*, 23–31. [CrossRef]
18. Nespolo, M.; Guillot, B. CHARDI2015: charge distribution analysis of non-molecular structures. *J. Appl. Cryst.* **2016**, *49*, 317–321. [CrossRef]
19. Adams, S.; Rao, R.P. Transport pathways for mobile ions in disordered solids from the analysis of energy-scaled bond-valence mismatch landscapes. *Phys. Chem. Chem. Phys.* **2009**, *11*, 3210–3216. [CrossRef]

20. Adams, S.; Rao, R.P. High power lithium ion battery materials by computational design. *Phys. Status Solidi* **2011**, *A208*, 1746–1753. [CrossRef]
21. Pauling, L. The principles determining the structure of complex ionic crystals. *J. Am. Chem. Soc.* **1929**, *51*, 1010–1026. [CrossRef]
22. Brown, I.D.; Altermatt, D. Bond-valence parameters obtained from a systematic analysis of the Inorganic Crystal Structure Database. *Acta Cryst.* **1985**, *B41*, 244–247. [CrossRef]
23. Adams, S. Relationship between bond valence and bond softness of alkali halides and chalcogenides. *Acta Cryst.* **2001**, *B57*, 278–287. [CrossRef] [PubMed]
24. Mazza, D. Modeling ionic conductivity in Nasicon structures. *J. Solid State Chem.* **2001**, *156*, 154–160. [CrossRef]
25. Marzouki, R.; Ben Smida, Y.; Guesmi, A.; Georges, S.; Ali, I.H.; Adams, S.; Zid, M.F. Structural and Electrical Investigation of New Melilite Compound $K_{0.86}Na_{1.14}CoP_2O_7$. *Int. J. Electrochem. Sci.* **2018**, *13*, 11648–11662. [CrossRef]
26. Ben Moussa, M.A.; Marzouki, R.; Brahmia, A.; Georges, S.; Obbade, S.; Zid, M.F. Synthesis and Structure of New Mixed Silver Cobalt (II)/(III) Diphosphate-$Ag_{3.68}Co_2(P_2O_7)_2$. Silver (I) Transport in the Crystal. *Int. J. Electrochem. Sci.* **2019**, *14*, 1500–1515. [CrossRef]
27. Chen, H.; Wong, L.L.; Adams, S. SoftBV–a software tool for screening the materials genome of inorganic fast ion conductors. *Acta Cryst.* **2019**, *B75*, 18–33. [CrossRef]
28. Ben Said, R.; Louati, B.; Guidara, K. Electrical properties and conduction mechanism in the sodium-nickel diphosphate. *Ionics* **2014**, *20*, 703–711. [CrossRef]
29. Li, H.; Zhang, Z.; Xu, M.; Bao, W.; Lai, Y.; Zhang, K.; Li, J. Triclinic Off-stoichiometric $Na_{3.12}Mn_{2.44}(P_2O_7)_2$/C Cathode Materials for High Energy/Power Sodium-ion Batteries. *Appl. Mater. Interfaces* **2018**, *10*, 24564–24572. [CrossRef]
30. Nakamoto, K. *Infrared and Raman Spectra of Inorganic and Coordination Compounds*, 3rd ed.; Wiley-Inter-Science: New York, NY, USA, 1978.
31. Krichen, M.; Gargouri, M.; Guidara, K.; Megdiche, M. Phase transition and electrical investigation in lithium copper pyrophosphate compound Li2CuP2O7 using impedance spectroscopy. *Ionics* **2017**, *23*, 3309–3322. [CrossRef]
32. Langlois, S.; Couret, F. Flow-through and flow-by porous electrodes of nickel foam. I. Material characterization. *J. Appl. Electrochem.* **1989**, *19*, 43. [CrossRef]
33. Hafidi, E.; El Omari, M.; El Omari, M.; Bentayeb, A.; Bennazha, J.; El Maadi, A.; Chehbouni, M. Conductivity studies of some diphosphates with the general formula AI2BIIP2O7 by impedance spectroscopy. *Arab. J. Chem.* **2013**, *6*, 253–263. [CrossRef]
34. Marzouki, R.; Guesmi, A.; Zid, M.F.; Driss, A. Synthesis, Crystal Structure and Electrical Properties of a New Mixed Compound $(Na_{0.71}Ag_{0.29})_2CoP_2O_7$. *Cryst. Struct. Theory Appl.* **2013**, *1*, 68–73. [CrossRef]
35. Ben Smida, Y.; Marzouki, R.; Guesmi, A.; Georges, S.; Zid, M.F. Synthesis, structural and electrical properties of a new cobalt arsenate $NaCo_2As_3O_{10}$. *J. Solid State Chem.* **2015**, *221*, 132–139. [CrossRef]
36. Marzouki, R.; Guesmi, A.; Zid, M.F.; Driss, A. Etude physico-chimiques du monoarséniate mixte $Na_4Co_{5.63}Al_{0.91}(AsO_4)_6$ et simulation des chemins de conduction. *Ann. Chim. Sci. Mat.* **2013**, *38*, 117. [CrossRef]
37. Nasri, R.; Marzouki, R.; Georges, S.; Obbade, S.; Zid, M.F. Synthesis, sintering, electrical properties, and sodium migration pathways of new lyonsite $Na_2Co_2(MoO_4)_3$. *Turk. J. Chem.* **2018**, *42*, 1251. [CrossRef]

© 2020 by the authors. Licensee MDPI, Basel, Switzerland. This article is an open access article distributed under the terms and conditions of the Creative Commons Attribution (CC BY) license (http://creativecommons.org/licenses/by/4.0/).

Article

Characterization of Poly(Ethylene Oxide) Nanofibers—Mutual Relations between Mean Diameter of Electrospun Nanofibers and Solution Characteristics †

Petr Filip * and Petra Peer

Institute of Hydrodynamics of the Czech Academy of Sciences, 16000 Prague, Czech Republic; peer@ih.cas.cz
* Correspondence: filip@ih.cas.cz
† This paper is an expanded version of "Electrospinning of poly(ethylene oxide) solutions—Quantitative relations between mean nanofibre diameter, concentration, molecular weight, and viscosity" published in Proceedings of Novel Trends in Rheology VIII, Zlin, Czech Republic, 30–31 July 2019.

Received: 11 November 2019; Accepted: 2 December 2019; Published: 12 December 2019

Abstract: The quality of electrospun poly(ethylene oxide) (PEO) nanofibrous mats are subject to a variety of input parameters. In this study, three parameters were chosen: molecular weight of PEO (100, 300, 600, and 1000 kg/mol), PEO concentration (in distilled water), and shear viscosity of PEO solution. Two relations free of any adjustable parameters were derived. The first, describing the initial stage of an electrospinning process expressing shear viscosity using PEO molecular weight and concentration. The second, expressing mean nanofiber diameter using concentration and PEO molecular weight. Based on these simple mathematical relations, it is possible to control the mean nanofiber diameter during an electrospinning process.

Keywords: electrospinning; poly(ethylene oxide); nanofiber diameter; molecular weight; concentration

1. Introduction

At present, nanofibrous mats are efficiently used in many applications: filters, tissue engineering, drug delivery systems, antibacterial wound dressing, protective clothing, nanocomposite materials, to name a few. One of the ways one can produce nanofibrous mats is through the process of electrospinning. In this process, polymer solution or melt is exposed to a high-voltage electric field (in orders of tens kV) under which viscoelastic polymer jets emanate from so-called Taylor cones [1] formed at the polymer surface. After passing approximately 10–20 cm in length, the material of the jets (after evaporation of a solvent) is cumulated on an earthed collector [2–4] at the shape of individual nanofibers forming a non-woven textile.

The problem is that some promising materials, such as chitosan, keratin, and other protein-based materials, cannot be electrospun in their pure forms. This contrasts with the easy spinnability [5,6] of, e.g., poly(ethylene oxide) (PEO). Fortunately, even the negligible presence of PEO (up to 2%) in solutions of the above-listed materials completely changes their disposition to being electrospun (see below). Intensive study of PEO behavior during the electrospinning process has been undertaken due to this fact and because of the excellent biodegradability, biocompatibility, and non-toxicity. These attributes also reflect PEO nanofiber applications in biomedicine and the food industry [7–10] apart from the above already mentioned improvements in spinnability in combination with chitin or chitosan [11–14], keratin [15,16], silk [17,18] and other materials.

As in any electrospun material the resulting properties of PEO nanofibrous mats are subject to four groups of entry parameters: the polymer (molecular weight, molecular weight distribution, topology of macromolecules), solvent (surface tension, solubility parameters, relative permittivity), solution (viscosity, concentration, specific conductivity), and process parameters (electric field strength, tip-to-collector distance, temperature, humidity). In no way can these parameters be analyzed separately as many of them are mutually interlaced as can be documented, for example, in the case of molecular weight, concentration, and viscosity.

The whole process of electrospinning is so complex that there is no possibility to express the behavior of one parameter—e.g., a mean nanofiber diameter—through the remaining ones. This can be documented by a relatively complicated process of nanofiber formation starting from an initial straightforward stable motion of liquid material towards a collector, consecutively changed to a 'whipping' (unstable, chaotic) motion, and finally, after solvent evaporation, converting to solid nanofibers. Therefore, there is a necessity to choose a limited number of crucial parameters that participate in the setting of a nanofiber diameter and fixing of the others.

Experimentally this approach was applied to determine the dependence of a mean nanofiber diameter on polymer concentration [19–28], on applied voltage [22–26,28,29], on viscosity [20,21], on solution flow rate [22,23,25,26,29], on tip-to-collector distance [22–24,30], on addition of salts [24,29], on the composition of the mixed solvent [31], on elasticity of the solution [32], on addition of nanoparticles [33,34], on addition of polyelectrolyte [35].

Electrospinning is a complex phenomenon to analyze because of the coupling between the electric field and the deformation of the fluid, the latter, in turn, determined by the rheology of the material [36]. The first theoretical approach in describing the process where electrospinning starts with a Taylor cone, passes through both regions (straightforward and spiraling) of viscoelastic jet, and then up to the deposition of nanofibers onto a collector, appeared approximately two decades ago. The beads–spring model (not to be confused with the appearance of unwanted beads violating the straight geometry of final nanofibers) was presented in [37]. A list of this and other models [38–46] describing the flow behavior of electrified and electrospun viscoelastic jets is introduced in [2].

The authors in [44] concluded that a final nanofiber diameter can be determined from knowledge of the flow rate, electric current, and the surface tension of the fluid. In [27], the authors present the independence of an initial straight jet diameter on solution concentration and indicate that a mean nanofiber diameter depends dominantly on the jet-whipping process. In [47], the authors applied the theoretical modeling in [42,43] and stated that the final diameter also depends on the initial diameter of a jet. They classify the importance of 13 entry parameters with respect to their influence on a mean nanofiber diameter and sort them into three groups: strong, moderate, and minor. The model involving solution viscosity, evaporation rate, and containing specific charge (electric current divided by the spinning throughput) was introduced in [48,49]. A prediction of a final nanofiber diameter published in [50] is based on the model published in [36]. However, the assumptions limit the model to relatively low polymer–solvent concentrations (<12 wt.%). A determination of the final nanofiber diameter considers solution flow rate, applied voltage, and polymer concentration.

Applicability of the electrospun nanofibrous mats (filters, membranes, etc.) is closely related to their permeability. Hence, the passage of the particles through the mats can also be controlled by molecular weight and concentration of a polymeric material used in the process of electrospinning. The standard experiments providing mean nanofiber diameters and shear viscosity dependent on molecular weights and used concentrations of polymeric material are time-consuming, with higher financial costs and cover only discrete values of entry parameters.

This contribution aims to propose the relations mutually relating three entry parameters (PEO molecular weight, PEO concentration, shear viscosity of PEO solution) and a resulting mean nanofiber diameter. In contrast to the experiments, these relations cover continuously and sufficiently broad regions of entry parameters creating the possibility of achieving a required mean nanofiber diameter. The proposed relations are compared with the experiments (four various PEO molecular weights), and the mean deviations are practically within the experimental errors.

2. Materials and Methods

2.1. Materials

Four various poly (ethylene oxide) (Sigma Aldrich, Saint Louis, MO, USA) batches differing in molecular weight (100, 300, 600, and 1000 kg/mol) were consecutively dissolved in distilled water at different concentration ranges (introduced in Table 1) meeting successful spinnability (subjected to a molecular weight) of the PEO solutions. The PEO solutions were prepared using a magnetic stirrer (Heidolph MR Hei-Tec, Schwabach, Germany) with the help of a Teflon-coated magnetic cross applied for 48 h under constant conditions (mixing rate 250 rpm and temperature 25 °C).

Table 1. A list of prepared poly(ethylene oxide) (PEO) solutions.

Molecular Weight (kg/mol)	Concentration (wt.%)
100	12, 15, 20, 22, 24, 28, 30, 34
300	5, 7, 8, 9, 10, 11, 12, 13, 15, 17
600	3, 5, 5.4, 6, 7, 8, 9
1000	3, 3.3, 3.75, 4, 5, 6

2.2. Rheological Characterization

Shear viscosity of the individual PEO solutions was determined at a constant temperature of 25 °C using a Physica MCR 501 device (Anton Paar, Graz, Austria) equipped with concentric cylinders (26.6/28.9 mm-inner/outer diameters). Each measurement was repeated at least three times with very good reproducibility.

2.3. The Electrospinning Process

Laboratory equipment consisting of a high voltage power supply (Spellman SL70PN150, Hauppauge, NY, USA), a carbon steel stick (10 mm in diameter) with a hollowed semi-spherical pit at the end for polymer filling, and equipped with a motionless flat metal collector (for details see [51]) was used for an electrospinning process. The drop of polymer solution filling the pit contains approximately 0.2 mL in volume, a tip-to-collector distance was fixed (20 cm) as well as nearly constant ambient conditions (temperature 23 ± 1 °C, relative humidity 41 ± 1%). Good quality of electrospun mats (elimination of appearance of web blobs) was ensured by a gradual voltage decrease from 25 kV fixed for M_w = 100 kg/mol to 12 kV fixed for M_w = 1000 kg/mol.

2.4. Nanofibrous Mat Characterization

A high-resolution scanning electron microscope (SEM) Vega 3 (Tescan, Brno, Czech Republic) was used for imaging of nanofibrous mats after their sputtering with a conductive layer to improve conductivity.

3. Results and Discussion

Due to a non-negligible number of entry parameters, the process of electrospinning cannot be simultaneously analyzed from all material, geometrical, and process aspects. It is always necessary to fix a substantial majority of entry parameters and concentrate on a moderate number of selected parameters. The impacts of the individual parameters are usually interlaced, and in the following analysis, we will pay attention to a mutual interplay between PEO molecular weight, PEO concentration, viscosity of a PEO solution in distilled water, and a mean diameter of the resulting nanofibers.

Specifically, the emphasis will focus on the derivation of two dependencies:

(1) A determination of functional relation between shear viscosity (η), PEO concentration (c), and PEO molecular weight (M_w);

(2) A determination of functional relation between the diameter of nanofibers (*dia*), PEO concentration (*c*), and PEO molecular weight (M_w).

From the viewpoint of easy and clear applicability of the proposed relations, they should exhibit the following attributes:

(a) Usage of elementary algebraic functions only;
(b) Absence of adjustable parameters;
(c) Their validity should cover sufficiently broad regions of two entry material characteristics—PEO molecular weight and PEO concentrations in distilled water (a range of concentrations shifts with respect to successful electrospinability of PEO solutions dependent on M_w);
(d) The correctness of the approximate relations-deviations of their predictions from the experimental data should potentially exceed the experimental errors only moderately.

The least accurate determination of a precise value out of the four studied parameters (M_w, c, η, *dia*) is represented by the mean nanofiber diameter *dia*.

A mean nanofiber diameter derived from 300 measurements taken from three different images was determined by applying the Adobe Creative Suite software (San Jose, CA, USA). Figure 1 displays SEM images of PEO nanofibers created from various molecular weights at different concentrations. The histograms attached to the individual molecular weights depict a variance of nanofiber diameters. This is also documented in Table 2.

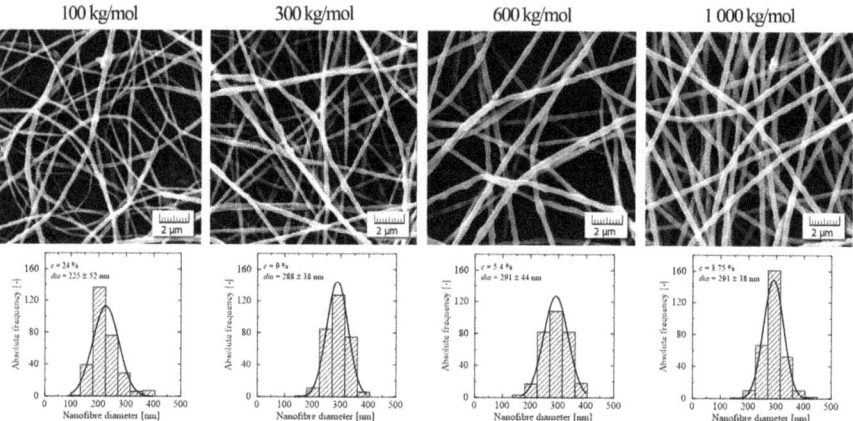

Figure 1. SEM pictures of nanofibrous mats for different poly(ethylene oxide) (PEO) solutions: M_w = 100 kg/mol, c = 24%; M_w = 300 kg/mol, c = 9%; M_w = 600 kg/mol, c = 5.4%; M_w = 1,000 kg/mol, c = 3.75%. The corresponding histograms describe the variance of nanofiber diameters.

Table 2. Variance of the measured nanofiber diameters.

M_w (kg/mol)	Dispersion of Nanofiber Diameters									
100	c (wt.%)	12	15	20	22	24	28	30	34		
	dia (nm)	113 ± 34	147 ± 33	174 ± 52	206 ± 49	225 ± 52	279 ± 56	303 ± 40	326 ± 43		
300	c (wt.%)	5	7	8	9	10	11	12	13	15	17
	dia (nm)	132 ± 26	210 ± 46	237 ± 44	288 ± 38	305 ± 48	326 ± 41	344 ± 47	367 ± 42	389 ± 53	442 ± 46
600	c (wt.%)	3	5	5.4	6	7	8	9			
	dia (nm)	141 ± 30	267 ± 42	291 ± 44	310 ± 44	331 ± 43	389 ± 28	430 ± 46			
1000	c (wt.%)	3	3.3	3.75	4	5	6				
	dia (nm)	206 ± 35	265 ± 29	291 ± 38	308 ± 49	391 ± 52	513 ± 75				

After analyzing and processing the experimental data obtained by consecutive electrospinning of PEO solutions differing in PEO molecular weight (100, 300, 600, and 1000 kg/mol) and concentration (see Table 2), we can calculate the correlation coefficients for the experimental data sets (*c*, *dia*). Their proximity to one (pure linearity behavior, see Table 3) indicates the possibility to approximate a mutual dependence of the mean nanofiber diameter on the concentration by a simple linear relation. To unify this approach, it is also necessary to express a general coefficient of linearity through the values of PEO molecular weights, which results in a slight deviation from the optimized values for the individual molecular weights.

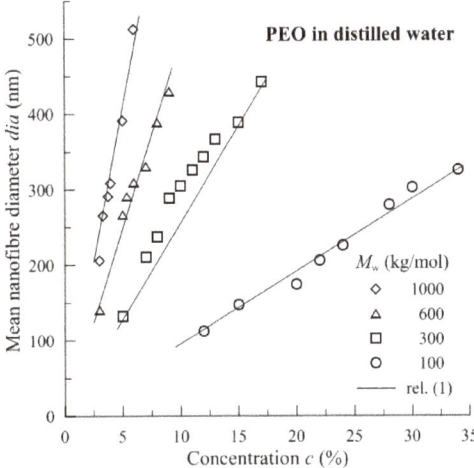

Figure 2. Dependence of mean nanofiber diameter (*dia*) on concentration (*c*) and molecular weight (M_w).

Table 3. Correlation coefficients between (*c*, *dia*) and (log(*c*), log(*η*)).

M_w (kg/mol)	Correlation Coefficient [-]	
	(*c*, *dia*) Figure 2	(log(*c*), log(*η*)) Figure 4
100	0.992	0.996
300	0.980	0.998
600	0.987	0.997
1000	0.992	0.942

Finally, we propose the following relation

$$dia = (a_1 M_w + a_2) \times c \quad (1)$$

with a linear proportionality between the mean nanofiber diameter and concentration, and with a linear dependence between the mean nanofiber diameter and molecular weight. The numerical values of the constants are $a_1 = 0.00008$ and $a_2 = 1.6$. The correspondence between the experimental and predicted data is depicted in Figure 2, with the mean deviation attaining 6.7%. Relation (1) complies with the tendencies introduced in literature [52,53], i.e., an increase in the nanofiber diameter both with increasing molecular weight and increasing concentration.

Dependence between shear viscosity, molecular weight, and concentration is a little more complicated. If we apply an analogous approach to that above, we obtain the correlation coefficients for the data sets (log(*c*), log(*η*)), again in close proximity to one (see Table 3). It justifies the proposal of a linear relation between log(*c*) and log(*η*). Unifying both a slope and an intercept with respect to the range of PEO molecular weights, a proposed relation is still algebraically simple.

$$\log(\eta) = b_1 \times \log(c) + [b_2 \times (\log(M_w))^2 + b_3 \times \log(M_w) + b_4], \quad (2)$$

where $b_1 = 4.71$, $b_2 = -0.82$, $b_3 = 12.7$, $b_4 = -48.8$. As can be seen, Relation (2) is composed of two separate members, the first one expressing a contribution of concentration only and the second one (in brackets) representing the participation of molecular weight exclusively. Figure 3 documents the courses of the proposed predicted curves attaining a mean deviation of 6.3% in the semi-log coordinates (concentration-linear, shear viscosity-logarithmic), in fully linear coordinates, a mean deviation attained 11.8%. In the log–log coordinates (Figure 4), there is a linear dependence of shear viscosity on PEO concentration with the fixed slope attaining a value of 4.71.

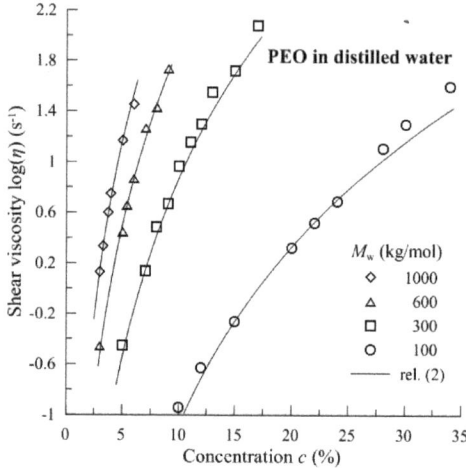

Figure 3. Dependence of shear viscosity ($\log(\eta)$) on concentration (c) and molecular weight (M_w) in semi-log coordinates.

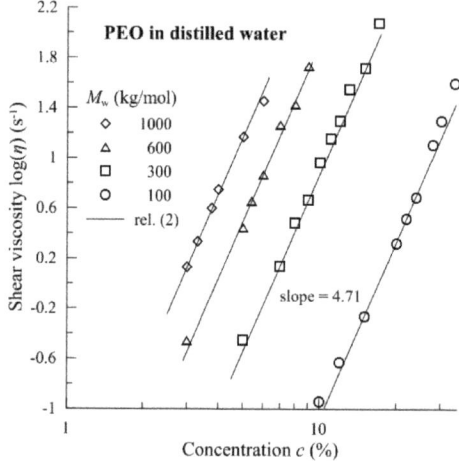

Figure 4. Dependence of shear viscosity ($\log(\eta)$) on concentration (c) and molecular weight (M_w) in log-log coordinates.

4. Conclusions

The introduced relationships propose a hint as to which way the geometrical arrangement of the resulting electrospun nanofibrous mats can be modified. Based on the relatively simple functional

Relation (1), it is possible to alter a mean nanofiber diameter by a suitable choice of PEO material (choice of an adequate molecular weight) and setting a corresponding concentration. Relation (2) characterizes the mutual interplay between three input parameters: molecular weight, concentration, and shear viscosity. Both Relations (1) and (2) are valid for the processed material-PEO-solved in distilled water. However, it can be expected that the analogous relations could also be derived for other combinations of polymers and solvents.

Author Contributions: Investigation, P.F. and P.P.; Writing, P.F.

Funding: This research was funded by the Grant Agency CR, grant number 17-26808S.

Conflicts of Interest: The authors declare no conflict of interest.

References

1. Taylor, G.I. Electrically driven jets. *Proc. R. Soc. A* **1969**, *313*, 453–475. [CrossRef]
2. Reneker, D.H.; Yarin, A.L. Electrospinning jets and polymer nanofibers. *Polymer* **2008**, *49*, 2387–2425. [CrossRef]
3. Bhardwaj, N.; Kundu, S.C. Electrospinning: A fascinating fiber fabrication technique. *Biotechnol. Adv.* **2010**, *28*, 325–347. [CrossRef] [PubMed]
4. Agarwal, S.; Greiner, A.; Wendorff, J.H. Functional materials by electrospinning of polymers. *Prog. Polym. Sci.* **2013**, *38*, 963–991. [CrossRef]
5. Deitzel, J.M.; Kleinmeyer, J.D.; Hirvonen, J.K.; Tan, N.C.B. Controlled deposition of electrospun poly (ethylene oxide) fibers. *Polymer* **2001**, *42*, 8163–8170. [CrossRef]
6. Son, W.K.; Youk, J.H.; Lee, T.S.; Park, W.H. The effects of solution properties and polyelectrolyte on electrospinning of ultrafine poly (ethylene oxide) fibers. *Polymer* **2004**, *45*, 2959–2966. [CrossRef]
7. Agarwal, S.; Wendorff, J.H.; Greiner, A. Use of electrospinning technique for biomedical applications. *Polymer* **2008**, *49*, 5603–5621. [CrossRef]
8. Rieger, K.A.; Birch, N.P.; Schiffman, J.D. Designing electrospun nanofiber mats to promote wound healing—A review. *J. Mater. Chem. B* **2013**, *1*, 4531–4541. [CrossRef]
9. Chew, S.Y.; Wen, Y.; Dzenis, Y.; Leong, K.W. The role of electrospinning in the emerging field of nanomedicine. *Curr. Pharm. Des.* **2006**, *12*, 4751–4770. [CrossRef]
10. Heunis, T.; Bshena, O.; Klumperman, B.; Dicks, L. Release of bacteriocins from nanofibers prepared with combinations of poly (D, L-lactide) (PDLLA) and poly (ethylene oxide) (PEO). *Int. J. Mol. Sci.* **2011**, *12*, 2158–2173. [CrossRef]
11. Jung, H.-S.; Kim, M.H.; Shin, J.Y.; Park, S.R.; Jung, J.-Y.; Park, W.H. Electrospinning and wound healing activity of β-chitin extracted from cuttlefish bone. *Carbohydr. Polym.* **2018**, *193*, 205–211. [CrossRef]
12. Kuntzler, S.G.; Vieira Costa, J.A.; de Morais, M.G. Development of electrospun nanofibers containing chitosan/PEO blend and phenolic compounds with antibacterial activity. *Int. J. Biol. Macromol.* **2018**, *117*, 800–806. [CrossRef] [PubMed]
13. Garcia, C.E.G.; Martínez, F.A.S.; Bossard, F.; Rinaudo, M. Biomaterials Based on Electrospun Chitosan. Relation between Processing Conditions and Mechanical Properties. *Polymers* **2018**, *10*, 257. [CrossRef] [PubMed]
14. Abid, S.; Hussain, T.; Nazir, A.; Zahir, A.; Ramakrishna, S.; Hameed, M.; Khenoussi, N. Enhanced antibacterial activity of PEO-chitosan nanofibers with potential application in burn infection management. *Int. J. Biol. Macromol.* **2019**, *135*, 1222–1236. [CrossRef] [PubMed]
15. Aluigi, A.; Vineis, C.; Varesano, A.; Mazzuchetti, G.; Ferrero, F.; Tonin, C. Structure and properties of keratin/PEO blend nanofibres. *Europ. Polym. J.* **2008**, *44*, 2465–2475. [CrossRef]
16. Ma, H.; Shen, J.; Cao, J.; Wang, D.; Yue, B.; Mao, Z.; Wu, W.; Zhang, H. Fabrication of wool keratin/polyethylene oxide nano-membrane from wool fabric waste. *J. Clean. Prod.* **2017**, *161*, 357–361. [CrossRef]
17. Jin, H.J.; Fridrikh, S.V.; Rutledge, G.C.; Kaplan, D.L. Electrospinning Bombyx mori silk with poly (ethylene oxide). *Biomacromolecules* **2002**, *3*, 1233–1239. [CrossRef]
18. Wharram, S.E.; Zhang, X.; Kaplan, D.L.; McCarthy, S.P. Electrospun silk material systems for wound healing. *Macromol. Biosci.* **2010**, *10*, 246–257. [CrossRef]

19. Deitzel, J.M.; Kleinmeyer, J.; Harris, D.; Tan, N.C.B. The effect of processing variables on the morphology of electrospun nanofibers and textiles. *Polymer* **2001**, *42*, 261–272. [CrossRef]
20. McKee, M.G.; Wilkes, G.L.; Colby, R.H.; Long, T.E. Correlations of Solution Rheology with Electrospun Fiber Formation of Linear and Branched Polyesters. *Macromolecules* **2004**, *37*, 1760–1767. [CrossRef]
21. Gupta, P.; Elkins, C.; Long, T.E.; Wilkes, G.L. Electrospinning of linear homopolymers of poly (methyl methacrylate): Exploring relationships between fiber formation, viscosity, molecular weight and concentration in a good solvent. *Polymer* **2005**, *46*, 4799–4810. [CrossRef]
22. Chowdhury, M.; Stylios, G.K. Analysis of the effect of experimental parameters on the morphology of electrospun polyethylene oxide nanofibres and on their thermal properties. *J. Text. Inst.* **2012**, *103*, 124–138. [CrossRef]
23. Kong, L.; Ziegler, G.R. Quantitative relationship between electrospinning parameters and starch fiber diameter. *Carbohydr. Polym.* **2013**, *92*, 1416–1422. [CrossRef] [PubMed]
24. Matabola, K.P.; Moutloali, R.M. The influence of electrospinning parameters on the morphology and diameter of poly (vinyledene fluoride) nanofibers-effect of sodium chloride. *J. Mater. Sci.* **2013**, *48*, 5475–5482. [CrossRef]
25. Ghorani, B.; Goswami, P.; Russell, S.J. Parametric Study of Electrospun Cellulose Acetate in Relation to Fibre Diameter. *Res. J. Text. Appar.* **2015**, *19*, 24–40. [CrossRef]
26. Şimşek, M.; Çakmak, S.; Gümüşderelioğlu, M. Insoluble poly (ethylene oxide) nanofibrous coating materials: Effects of crosslinking conditions on the matrix stability. *J. Polym. Res.* **2016**, *23*, 236. [CrossRef]
27. Wang, C.; Wang, Y.; Hashimoto, T. Impact of entanglement density on solution electrospinning: A phenomenological model for fiber diameter. *Macromolecules* **2016**, *49*, 7985–7996. [CrossRef]
28. Lasprilla-Botero, J.; Alvarez-Lainez, M.; Lagaron, J.M. The influence of electrospinning parameters and solvent selection on the morphology and diameter of polyimide nanofibers. *Mater. Today Commun.* **2018**, *14*, 1–9. [CrossRef]
29. Zong, X.; Kim, K.; Fang, D.; Ran, S.; Hsiao, B.S.; Chu, B. Structure and process relationship of electrospun bioabsorbable nanofiber membranes. *Polymer* **2002**, *43*, 4403–4412. [CrossRef]
30. Long, F.C.; Kamsom, R.A.; Nurfaizey, A.H.; Isa, M.H.M.; Masripan, N.A.B. The influence of electrospinning distances on fibre diameter of poly (vinyl alcohol) electrospun nanofibers. In Proceedings of the 4th Mechanical Engineering Research Day (MERD), University Teknikal Malaysia Melaka, Melaka, Malaysia, 30 March 2017; Bin Abdollah, M.F., Tuan, T.B., Salim, M.A., Akop, M.Z., Ismail, R., Musa, H., Eds.; pp. 377–378.
31. Han, S.O.; Youk, J.H.; Min, K.D.; Kang, Y.O.; Park, W.H. Electrospinning of cellulose acetate nanofibers using a mixed solvent of acetic acid/water: Effects of solvent composition on fiber diameter. *Mater. Lett.* **2008**, *62*, 759–762. [CrossRef]
32. Gupta, D.; Jassal, M.; Agrawal, A.K. Electrospinning of poly (vinyl alcohol)-based Boger fluids to understand the role of elasticity on morphology of nanofibres. *Ind. Eng. Chem. Res.* **2015**, *54*, 1547–1554. [CrossRef]
33. Sundarrajan, S.; Ramakrishna, S. Fabrication of nanocomposite membranes from nanofibers and nanoparticles for protection against chemical warfare stimulants. *J. Mater. Sci.* **2007**, *42*, 8400–8407. [CrossRef]
34. Yang, Q.B.; Li, D.M.; Hong, Y.L.; Li, Z.Y.; Wang, C.; Qiu, S.L.; Wei, Y. Preparation and characterization of a PAN nanofiber containing Ag nanoparticles via electrospinning. *Synth. Metals* **2003**, *137*, 973–974. [CrossRef]
35. Sundarrajan, S.; Venkatesan, A.; Ramakrishna, S. Fabrication of nanostructured self-detoxifying nanofiber membranes that contain active polymeric functional groups. *Macromol. Rapid Commun.* **2009**, *30*, 1769–1774. [CrossRef] [PubMed]
36. Feng, J.J. The stretching of an electrified non-Newtonian jet: A model for electrospinning. *Phys. Fluids* **2002**, *14*, 3912–3926. [CrossRef]
37. Yarin, A.L.; Koombhongse, S.; Reneker, D.H. Bending instability in electrospinning of nanofibers. *J. Appl. Phys.* **2001**, *89*, 3018–3026. [CrossRef]
38. Spivak, A.F.; Dzenis, Y.A. Asymptotic decay of radius of a weakly conductive viscous jet in an external electric field. *Appl. Phys. Lett.* **1998**, *73*, 30673069. [CrossRef]
39. Hohman, M.M.; Shin, M.; Rutledge, G.; Brenne, R.M.P. Electrospinning and electrically forced jets. I. Stability theory. *Phys. Fluids* **2001**, *13*, 2201–2220. [CrossRef]
40. Hohman, M.M.; Shin, M.; Rutledge, G.; Brenner, M.P. Electrospinning and electrically forced jets II. Applications. *Phys. Fluids* **2001**, *13*, 2221–2236. [CrossRef]

41. Feng, J.J. Stretching of a straight electrically charged viscoelastic jet. *J. Non-Newton. Fluid Mech.* **2003**, *116*, 55–70. [CrossRef]
42. Reneker, D.H.; Yarin, A.L.; Fong, H.; Koombhongse, S. Bending instability of electrically charged liquid jets of polymer solution in electrospinning. *J. Appl. Phys.* **2000**, *87*, 4531–4547. [CrossRef]
43. Yarin, A.L.; Koombhongse, S.; Reneker, D.H. Taylor cone and jetting from liquid droplets in electrospinning of nanofibers. *J. Appl. Phys.* **2001**, *90*, 4836–4846. [CrossRef]
44. Fridrikh, S.V.; Yu, J.H.; Brenner, M.P.; Rutledge, G.C. Controlling the fiber diameter during electrospinning. *Phys. Rev. Lett.* **2003**, *90*, 144502. [CrossRef] [PubMed]
45. Theron, S.A.; Yarin, A.L.; Zussman, E.; Kroll, E. Multiple jets in electrospinning: Experiment and modeling. *Polymer* **2005**, *46*, 2889–21899. [CrossRef]
46. Reneker, D.H.; Yarin, A.L.; Zussman, E.; Xu, H. Electrospinning of nanofibers from polymer solutions and melts. *Adv. Appl. Mech.* **2007**, *41*, 43–195.
47. Thompson, C.J.; Chase, G.G.; Yarin, A.L.; Reneker, D.H. Effects of parameters on nanofiber diameter determined from electrospinning model. *Polymer* **2007**, *48*, 6913–6922. [CrossRef]
48. Stepanyan, R.; Subbotin, A.; Cuperus, L.; Boonen, P.; Dorschu, M.; Oosterlinck, F.; Bulters, M.J.H. Fiber diameter control in electrospinning. *Appl. Phys. Lett.* **2014**, *105*, 173105–173109. [CrossRef]
49. Stepanyan, R.; Subbotin, A.; Cuperus, L.; Boonen, P.; Dorschu, M.; Oosterlinck, F.; Bulters, M.J.H. Nanofiber diameter in electrospinning of polymer solutions: Model and experiment. *Polymer* **2016**, *97*, 428–439. [CrossRef]
50. Ismail, N.; Maksoud, F., Jr.; Ghaddar, N.; Ghali, K.; Tehrani-Bagha, A. Simplified modeling of the electrospinning process from the stable jet region to the unstable region for predicting the final nanofiber diameter. *J. Appl. Polym. Sci.* **2016**, *133*, 44112. [CrossRef]
51. Peer, P.; Stenicka, M.; Pavlinek, V.; Filip, P. The storage stability of polyvinylbutyral solutions from an electrospinnability standpoint. *Polym. Degrad. Stab.* **2014**, *105*, 134–139. [CrossRef]
52. Koski, A.; Yim, K.; Shivkumar, S. Effect of molecular weight on fibrous PVA produced by electrospinning. *Mater. Lett.* **2004**, *58*, 493–497.
53. Shenoy, S.L.; Bates, W.D.; Frisch, H.L.; Wnek, G.E. Role of chain entanglements on fiber formation during electrospinning of polymer solutions: Good solvent, nonspecific polymer-polymer interaction limit. *Polymer* **2005**, *46*, 3372–3384.

© 2019 by the authors. Licensee MDPI, Basel, Switzerland. This article is an open access article distributed under the terms and conditions of the Creative Commons Attribution (CC BY) license (http://creativecommons.org/licenses/by/4.0/).

Review

Metal–Organic Framework Thin Films: Fabrication, Modification, and Patterning

Yujing Zhang and Chih-Hung Chang *

School of Chemical, Biological and Environmental Engineering, Oregon State University, Corvallis 97330, OR, USA; zhangyuj@oregonstate.edu
* Correspondence: Chih-Hung.Chang@oregonstate.edu

Received: 10 February 2020; Accepted: 19 March 2020; Published: 24 March 2020

Abstract: Metal–organic frameworks (MOFs) have been of great interest for their outstanding properties, such as large surface area, low density, tunable pore size and functionality, excellent structural flexibility, and good chemical stability. A significant advancement in the preparation of MOF thin films according to the needs of a variety of applications has been achieved in the past decades. Yet there is still high demand in advancing the understanding of the processes to realize more scalable, controllable, and greener synthesis. This review provides a summary of the current progress on the manufacturing of MOF thin films, including the various thin-film deposition processes, the approaches to modify the MOF structure and pore functionality, and the means to prepare patterned MOF thin films. The suitability of different synthesis techniques under various processing environments is analyzed. Finally, we discuss opportunities for future development in the manufacturing of MOF thin films.

Keywords: metal–organic framework; thin film; fabrication; patterning

1. Introduction

Metal–organic frameworks (MOFs) are a class of inorganic–organic hybrid crystalline microporous materials consisting of a highly ordered array of metal cations connected by multidentate organic linkers. The regular and extended network built by metal ions (or clusters) and organic linkers often forms a repeating cage-like structure, which grants MOFs an extensive internal surface area. In contrast to other porous materials, MOFs also possess designable structures that can be engineered with tailored pores for selective adsorption of specific gases [1]. Moreover, MOFs show outstanding features as in structural flexibility, thermal and chemical stability, etc., which grant MOFs great potential in numerous applications, such as gas storage and separation [2–5], liquid purification [6–9], catalysis [10–13], gas/chemical sensing [14–18], and energy production [19–22]. Other than direct applications, MOFs have also been used as precursors/templates for the production of inorganic functional materials with unique designability [23]. According to the needs in various applications, thousands of MOFs have been synthesized by now since being reported in the 1990s [1,24–26]. Nowadays, MOFs are available in various structures, such as nanocrystals (NCs) [27], nanospheres [28], nanosheets [29], needles [30], hierarchical monoliths [31], thin films (TFs) [32], membranes [33], and glasses [34–36]. Among these structures, MOF-TFs are drawing increasing attention due to their tremendous potential in the development of nanotechnology-enabling applications, such as optics [37], photonics [38], electronics [39], catalytic coatings [40], sensing [41–44], solar cell [45], battery [46], and supercapacitor [44]. One thing to notice is that MOF-TFs cannot be differentiated from MOF membranes by their chemical composition or by their selection of substrates. Concerning the definition of a membrane [47], which should be a medium that allows transfer to occur under a certain driving force, a TF [48] does not have such a restriction. The main difference between TFs and membranes lies in their functions, although there is reporting of freestanding MOF-TFs that can function as a membrane due to its porosity [49].

A significant amount of research is devoted to the fabrication, characterization, and application of MOF-TFs. MOF-TFs deposited on substrates of various functions enable different applications. For example, those on quartz crystal microbalance (QCM) substrates allow for the study of adsorption within MOF layers [50–52], those on gold substrates realize surface plasmon resonance (SPR) spectroscopy [53], and those on conducting electrodes open up a way for electrical and electrochemical applications [54,55]. The selection of substrate is crucial in the fabrication of MOF-TFs, especially in the deposition process based on hydro/solvothermal mother solution synthesis that requires good heterogeneous nucleation and growth of MOFs. To date, the fabrication of MOF-TFs has been realized on various substrates, such as nonplanar substrates [56], planar substrates [57], flexible substrates [42,58], and substrates terminated with different functional groups on the surface [59,60]. In spite of these extensive studies, engineering MOF-TFs with a controllable thickness and structure and precise chemical composition has always been a challenge. Commercialization involving scale-up production with an effective cost is an additional challenge [61]. As a result, investigation regarding novel synthesis approaches for the manufacturing of MOF-TFs continues.

Since there are many excellent reviews [33,62–66] focusing on the fabrication of MOF membranes, including the concept of mixed matrix membranes [67,68], this review covers current progress on the manufacturing of MOF-TFs. It summarizes and analyzes various fabrication processes for MOF-TFs, approaches for the modification of architectures and pore functionality in MOF-TFs, means for the preparation of patterned MOF-TFs, and application of different synthesis strategies under various processing environments. This review aims to advance the understanding of the processing of MOF-TFs to realize a more controllable, scalable, and eco-friendly synthesis. Concerning these objectives, some prospects on future opportunities for the development of the manufacturing of MOF-TFs are discussed at the end.

2. General Fabrication Techniques for MOF-TFs

Several excellent reviews focusing on MOFs, including MOF-TFs, have been published [37,69–77]. These papers review existing and potential applications, as well as the synthesis methods of MOF-TFs. However, regarding the fabrication techniques, these reviews focus on more commonly used approaches, such as hydro/solvothermal synthesis, the stepwise layer-by-layer (LBL) deposition method, and the electrochemical method. Considering the extensive discussion about the diverse synthesis of MOFs in other structures, such as nanoparticles (NPs) [78], composite structures [79], and membranes [33], here we focus on the fabrication of MOF-TFs, giving a comprehensive review on various synthesis strategies that are reported for MOF-TFs (including patterns) and offering a direction for future development for the green fabrication of MOF-TFs.

Classification of Fabrication Techniques

The classification of fabrication techniques in this review is according to the phase of the precursors in the synthesis reaction, such as liquid, solid, vapor, and gel. In the section of general fabrication techniques for MOF-TFs, we will focus the discussion on the processes using two precursors, one of which is the metal precursor and the other the organic precursor. The studies that require more than one metal or organic species will be discussed in the section of modification of MOF-TFs, with an emphasis on the manufacturing of TF structure/composition with modified functionality.

The discussion on general fabrication techniques will be grouped into three sections, including liquid–liquid synthesis methods, liquid–solid synthesis methods, and all other types of synthesis methods (i.e., solid–solid synthesis methods, vapor–solid synthesis methods, vapor–gel synthesis methods, and the post-assembly method).

3. General Liquid–Liquid Synthesis

Most syntheses of MOF-TFs are carried out via liquid-phase reaction, where both the metal and the organic precursors are dissolved in a solvent before the reaction. Typically, the dissolved precursors are either well mixed to prepare a solution mixture as the mother solution before the addition of substrates (such as in a direct solvothermal synthesis) or used separately in contact with the substrates in a sequential manner (such as in a stepwise LBL deposition). The high concentration of ionized reactants (i.e., the metal cations and the deprotonated organic linkers) in the solution leads typically to a homogeneous reaction in the solvent phase and simultaneously a heterogeneous reaction at the substrate surface [80]. There is competition between these two reactions, where the homogeneous reaction results in the formation of MOF crystals in the solvent while the heterogeneous reaction brings about the formation of MOF-TFs on the substrate surface. Therefore, promoting the heterogeneous reaction and suppressing the homogeneous reaction is a promising solution to realize the green and economical synthesis of high-quality MOF-TFs with good uniformity and continuity.

3.1. Direct Synthesis

Although numerous fabrication techniques have been developed for MOF-TFs, a one-pot hydro/solvothermal batch synthesis strategy is still the primary way and serves as the foundation for the development of many other synthesis strategies. Hydrothermal synthesis refers to those that happen in an aqueous solution above the boiling point of water while solvothermal synthesis is in a non-aqueous solution at relatively high temperatures [81]. Both types of synthesis often proceed in a sealed reactor, such as an autoclave, and a pressure vessel. In a typical hydro/solvothermal synthesis, the substrate is placed in a mixture of precursor solutions for MOFs and is subjected to a reaction at a high temperature. This one-pot hydro/solvothermal synthesis strategy for MOF-TFs can be categorized into two subcategories, i.e., the direct growth on unmodified substrates, and the secondary growth that involves the preparation of functional substrate surfaces. The main issues in the direct synthesis are the lack of control in preparing a homogeneous TF and possible substrate corrosion.

Based on the concept of direct synthesis, Cui et al. [82] realized the fabrication of MOF-TFs on rough surfaces via an in situ hydrothermal synthesis strategy. Stainless steel wires were used as the substrates, which were etched by hydrofluoric acid before the reaction since a rough surface can improve the retention of particles and thereby promotes the heterogeneous nucleation and growth of MOFs. After 8 h of reaction at 120 °C, complete coverage of HKUST-1 ($Cu_3(BTC)_2(H_2O)_3$, where BTC = 1,3,5-benzenetricarboxylate; also known as MOF-199) TF with a thickness of about 40 μm was obtained on an etched SSW (Figure 1A). The resulting HKUST-1-coated SSWs were tested for solid-phase microextraction for volatile and harmful benzene homologues, acquiring a low limit of detection that was 8.3–23.3 ng/L. Later, Sheberla et al. [83] reported the fabrication of MOF-TFs on smooth surfaces via an in situ solvothermal synthesis strategy. $Ni_3(HITP)_2$ MOF-TFs, where HITP = 2,3,6,7,10,11-hexaiminotriphenylene, were fabricated directly on quartz substrates under the reaction conditions in this study and a microporous structure and ultrahigh electrical conductivity was obtained that was of interest in electronic devices. The conductivity of the resulting $Ni_3(HITP)_2$ MOF-TFs reached 40 S/cm, which stood for the best records for MOFs and coordination polymers at the time. Campbell et al. [84] achieved continuous and dense Mg-MOF-74 (Mg_2DOBDC, where DOBDC = 2,5-dihydroxyterephthalate) TFs directly on porous alumina (Al_2O_3) substrates via an in situ solvothermal synthesis strategy (Figure 1B). In this report, the process was optimized by tuning the reaction conditions (i.e., reaction time, the dose of precursors, and the composition of solvents) that could influence the TF formation process and resulting thickness and morphology. A proper solvent composition of N,N-dimethylformamide (DMF):water:ethanol was reported, 16:2:2, in order to obtain well-intergrown Mg-MOF-74 TFs. In this study, it was noticed that, by fixing all the other reaction parameters, the film thickness increased from 1.6 to 1.8 μm when increasing the reaction time from 2.5 to 6 h. The results corresponded to a decreased growth rate that was from 10.67 to 5.00 nm/minute when increasing the reaction time, which implies a common disadvantage for solvothermal synthesis. Many types of MOF-TFs, such as MOF-5 ($Zn_4O(BDC)_3$, where BDC = 1,4-benzodicarboxylate; as known as IRMOF-1) [85], PCN-221 ($Zr_8(\mu_4\text{-}O)_6(OH)_8(TCPP)_3$, where TCPP = tetrakis(4-carboxylatephenyl)porphyrin), PCN-222 ($Zr_6(\mu_3\text{-}OH)_8(OH)_8(TCPP)_2$), and PCN-223 ($Zr_6(\mu_3\text{-}O)_4(\mu_3\text{-}OH)_4(TCPP)_3$) [86], can be achieved via direct synthesis.

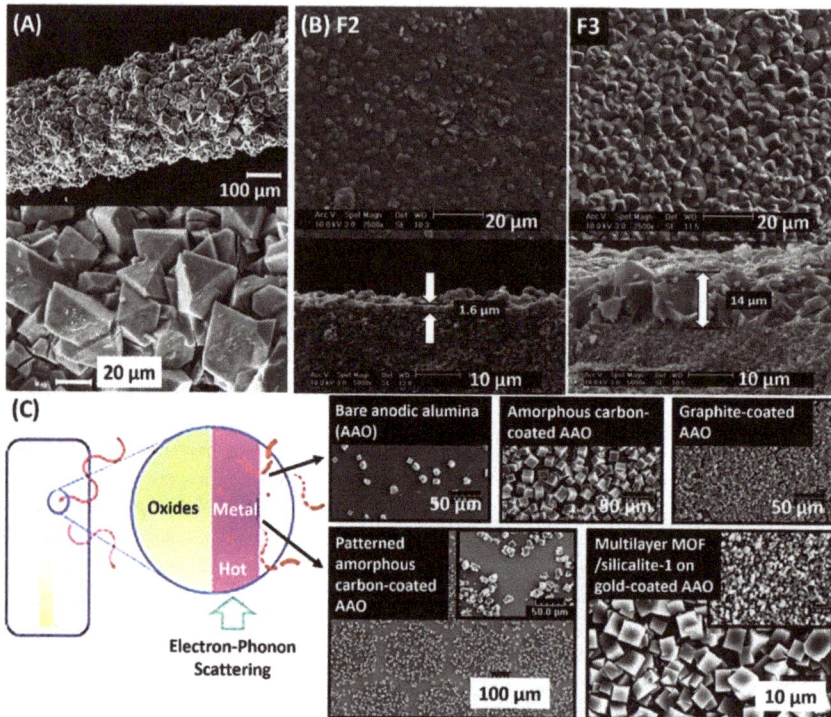

Figure 1. (**A**) SEM images of HKUST-1-coated stainless steel wire at different magnifications. (Reproduced with permission from Cui et al., Analytical Chemistry; published by American Chemical Society, 2009.) (**B**) SEM images of the surface and the cross-section of Mg-MOF-74 thin films formed using different precursor solutions. (Reproduced with permission from Campbell et al., Microporous and Mesoporous Materials; published by Elsevier BV, 2017.) (**C**) Schematic of the microwave-induced solvothermal synthesis, and SEM images of MOF-5 grown on different substrates after 30 s of the microwave-induced solvothermal reaction. (Reproduced with permission from Yoo et al., Chemical Communications; published by Royal Society of Chemistry, 2008.)

Different from the traditional direct synthesis based on the one-pot hydro/solvothermal synthesis, microwave irradiation was introduced into the reaction scheme by expediting the reaction of MOF-TFs because crystal growth is generally faster under microwave irradiation. With this consideration, Yoo and Jeong [56] reported a rapid synthesis of MOF-5 TFs on various surfaces that was finished in 5–30 s under microwave irradiation with a power of 500 W (Figure 1C). Uncoated, amorphous carbon-coated, graphite-coated, and gold-coated Al_2O_3 supports were investigated as substrates. The results illustrated an advantage of substrate surface modification in the fabrication of MOF-TFs (including patterns) that we will discuss in the secondary growth section. Furthermore, Bux et al. [87] reported that the microwave-assisted fabrication of MOF-TFs does not necessarily require a chemically modified substrate surface, nor an atomically smooth surface with electrostatic compatibility. In this study, dense and continuous ZIF-8 ($Zn(2\text{-mIm})_2$, where 2-mIm = 2-methylimidazole) TFs were acquired directly on porous titanium oxide substrates via a microwave-assisted solvothermal synthesis after

4 h of reaction at 100 °C. This type of synthesis strategy has also succeeded in the fabrication of other types of MOF-TFs on various substrates [88], demonstrating a great potential of microwave irradiation in the development of rapid synthesis strategies.

3.2. Secondary Growth

Although traditional hydro/solvothermal synthesis in the fabrication of MOF-TFs has been successful, it is still challenging to fabricate continuous MOF-TFs on unmodified substrates because the heterogeneous nucleation of MOF crystals on substrate surfaces is commonly inefficient [89]. Moreover, it is difficult for MOF crystals that have formed in solution to adhere to substrate surfaces due to the lack of binding sites. One of the main strategies to improve the heterogeneous nucleation and growth of MOF-TFs is through substrate surface modification [90]. This type of synthesis strategy is classified as secondary growth, which commonly involves the functionalization of the substrate by adding a functional layer on the surface before the synthesis. Other than traditional one-pot hydro/solvothermal synthesis, the secondary growth is often conducted with other synthesis strategies to achieve controllable formation of MOF-TFs [91].

Under the concept of secondary growth, functional group-terminated surfaces have been widely studied to improve the heterogeneous nucleation of MOF-TFs [92], as they can help to anchor the metal/metal-oxo nodes and/or organic linkers on substrate surfaces. Many kinds of functional layers are used for substrate surface modification, which can be grouped into two categories, including the organic functional layers, such as self-assembled monolayers (SAMs) and polymers, and the inorganic-involving functional layers, such as MOF NCs and metal oxide NPs.

SAMs are a class of organic functional layers used extensively in the fabrication of MOF-TFs. Commonly, SAMs consist of organothiol-based chains prepared on solid surfaces [93] to coordinate with MOF precursors. By changing the type of functional group on organothiol-based chains or the density, the growth orientation of MOF-TFs can be systematically controlled [94], contributing to the fabrication of highly oriented MOF-TFs. Hermes et al. [59] achieved the patterning of MOF-5 TFs based on patterned COOH-/CF$_3$-terminated SAMs prepared by microcontact printing (µCP) on gold surfaces. In this study, a mixture of Zn(NO$_3$)$_2$ and terephthalic acid was prepared in pure dimethylformamide at 75 °C as the mother solution for MOF-5. After mixing for 72 h, the solution was heated to 105 °C and rapidly cooled down to 25 °C to allow the crystallization of MOF-5. A clear supersaturated reaction mixture was obtained by filtration. Then, the µCP-patterned SAMs of 16-mercaptohexadecanoic acid (MHDA) and 1H,1H,2H,2H-perfluorododecane thiol (PFDT) on Au(111) were immersed in the reaction mixture, resulting in the selective deposition of MOF-5 on patterned areas. Biemmi et al. [94] reported a study on the influence of the type of SAMs on the orientation of the resulting MOF-TFs (Figure 2A). Based on the different coordination of –OH and –COOH groups, the resulting HKUST-1 TFs showed preferred [111] and [100] growth directions on the gold surfaces, respectively. A controlled orientation is beneficial in MOFs because it influences the pore system in MOF-TFs, which could open the way for more advanced applications based on improved adsorption [94]. Liu et al. [95] discovered that even by the same functional group-terminated SAMs, the orientation of the resulting HKUST-1 TFs could be different by varying the density of the functional groups. Zacher et al. [96] achieved highly oriented HKUST-1 TFs on bare Al$_2$O$_3$ surfaces and SAMs prepared on silicon dioxide (SiO$_2$) surfaces (Figure 2B). In the study, it was observed that there was no crystal nucleation on the bare SiO$_2$ surface based on oxidized Si wafer while densely packed polycrystalline agglomerated HKUST-1 microcrystals formed on sapphire (Al$_2$O$_3$) and atomic layer deposition (ALD)-Al$_2$O$_3$ surfaces. It was stated that because of electrostatic effct the nucleation of HKUST-1 preferred the basic surface of Al$_2$O$_3$ rather than the acidic surface of SiO$_2$. Therefore, the alkaline environment in the reaction system would be another key aspect to consider during experiment design depending on the type of MOFs. Other than MOF-5 and HKUST-1 TFs, different kinds of MOF-TFs are achieved by similar methods, such as Fe-MIL-88B-NH$_2$ (Fe$_3$O(BDC-NH$_2$)$_3$Cl, where BDC-NH$_2$ = 2-amino-1,4-benzenedicarboxylate) and its isomer Fe-MIL-101-NH$_2$ [60], and CAU-1 (Al$_4$(OH)$_2$(OCH$_3$)$_4$(BDC-NH$_2$)$_3$) [97].

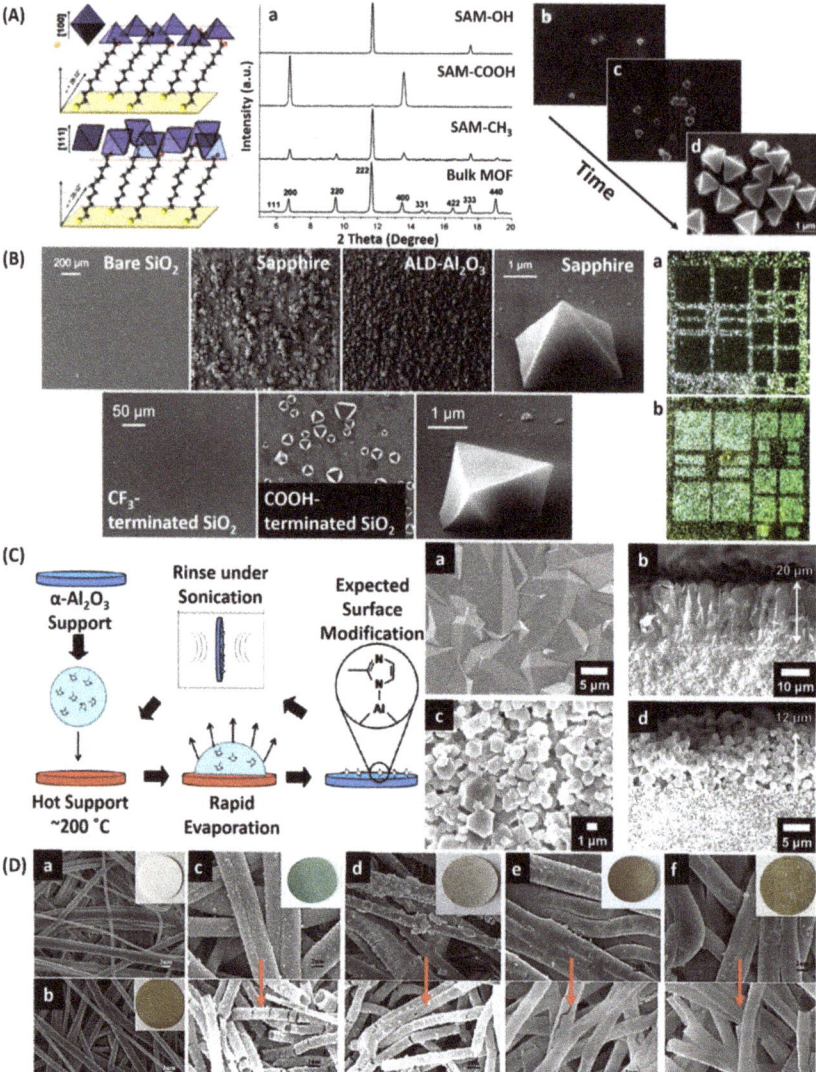

Figure 2. (**A**) Schematic illustrations of the oriented growth of HKUST-1 nanocrystals controlled via SAMs. (**a**) XRD patterns of HKUST-1 TFs on functionalized gold surfaces, compared with a randomly oriented HKUST-1 bulk sample measurement, and SEM images of HKUST-1 crystals on OH-terminated SAMs after immersion in the mother solution for (**b**) 16, (**c**) 24, and (**d**) 45 h. All scale bars, 1 µm. (Reproduced with permission from Biemmi et al., Journal of the American Chemical Society; published by American Chemical Society, 2007.) (**B**) SEM images of HKUST-1 TFs on different surfaces, a single pyramidal crystal grown on c-plane sapphire, and a single octahedral crystal grown on COOH-terminated Si/SiO$_2$. Optical images of HKUST-1 TFs on (**a**) a "positive" CF$_3$/COOH and (**b**) a "negative" COOH/CF$_3$ patterned SAM surface. (Reproduced with permission from Zacher et al., Journal of Materials Chemistry; published by Royal Society of Chemistry, 2007.) (**C**) Illustration of the substrate modification process. (**a,c**) Top-view and (**b,d**) cross-section SEM of a well-intergrown and a continuous but poorly intergrown ZIF-8 TF, respectively. (Reproduced with permission from McCarthy et al., Langmuir; published by American Chemical Society, 2010.) (**D**) SEM images of (**a**) an original

polypropylene (PP) fibrous membrane and (**b**) a polydopamine (PDA)-coated PP membrane; all scale bars, 3 µm. SEM images of (**c**) HKUST-1-, (**d**) MOF-5-, (**e**) MIL-100(Fe)-, and (**f**) ZIF-8-coated PDA-modified PP membranes (all scale bars, 2 µm), and the corresponding HKUST-1, MOF-5, MIL-100(Fe), and ZIF-8 nanotubes after the removal of the underlying PP fibers (all scale bars, 1 µm). Inserted are the corresponding optical photos of samples. (Reproduced with permission from Zhou et al., Chemical Communications; published by Royal Society of Chemistry, 2015.)

Similar to SAM-assisted synthesis, there is one type of surface modification based on the organic linkers of the desired MOF. McCarthy et al. [98] demonstrated the effectiveness of functionalizing α-Al_2O_3 substrates with organic compounds, which was benzimidazole (bIm) for ZIF-7 and 2-mIm for ZIF-8, via a rapid evaporation process (Figure 2C). This simple surface modification procedure provided strong covalent bonds between the α-Al_2O_3 substrates and the imidazolate linkers, which is effective in promoting the heterogeneous nucleation and growth of MOF-TFs. Take the synthesis of well-intergrown ZIF-8 in this study, for example. For substrate preparation, polished α-Al_2O_3 substrates were dried in a convection oven at 200 °C for 2 h. Then, 0.5–1 mL if methanolic solution of 2-mIm was dropped on a 2.2-cm^2 substrate. After it was dried, the substrate was removed from the oven and sonicated in methanol for about 30 s. A thoroughly-modified substrate was prepared by repeating this process about six times. The formation of ZIF-8 films on the modified substrate was finished via traditional solvothermal treatment in the ZIF-8 mother solution. After 4 h of reaction at 120 °C, well-intergrown ZIF-8 film was obtained on the porous α-Al_2O_3 substrate, which showed high selectivities of 11.6 and 13 for H_2/N_2 and H_2/CH_4, respectively.

Polymers can serve as nucleation centers for MOFs as well. Zhou et al. [99] investigated the fabrication of different MOF-TFs based on polydopamine (PDA)-coated substrates. HKUST-1, MOF-5, MIL-100(Fe) ($Fe_3O(H_2O)_2OH(BTC)_2$), and ZIF-8 TFs were successfully deposited on PDA-modified substrates in an LBL deposition manner (Figure 2D). In this study, PDA functioned as an effective nucleation center on the fibers for MOFs because the catechol group on PDA has a strong coordination ability with metal cations. Hence, the heterogeneous nucleation and growth of MOF-TFs were improved through the coordination of the metal cations and the catechol group. Compared to uncoated polypropylene (PP) fibrous membrane, the MOF-modified PP showed excellent adsorption in terms of efficient removal of rhodamine B from water. Complete removal of rhodamine B based on an MIL-100-PP membrane was achieved in 3 h at 40 °C, while there was barely any removal based on a bare PP membrane under the same testing conditions.

Inorganic seed-assisted secondary growth relies on preformed nano-sized metal-based seeds to assist the nucleation of MOF-TFs. One type of inorganic-involving seed is preformed MOF NCs. Bux et al. [100] fabricated a highly oriented ZIF-8 film via seed-assisted secondary growth. ZIF-8 NCs were prepared on a porous α-Al_2O_3 substrate via the hydrogen bonds formed with polyethyleneimine, which worked as the coupling agent between the ZIF-8 seeds and Al_2O_3 surface. The ZIF-8-seeded substrate was obtained after it was immersed in seeding solution using an automatic dip-coating device with defined dipping and withdrawing speeds, followed by traditional solvothermal synthesis to achieve continuous and well-intergrown ZIF-8 TF. XRD analysis of the resulting film showed a preferred orientation at the [100] direction parallel to the support that was explained by an evolutionary growth process. The resulting mesoporous and microporous structure showed excellent performance in H_2/C_3H_8 separation, with a separation factor above 300. Papporello et al. [101] demonstrated the fabrication of ZIF-8 TFs on copper-based substrates, in which the commercially available ZIF-8 NCs were attached to copper foils as seeds via a manual rubbing manner (Figure 3A). It was observed that when using methanol as the precursor solvent, the presence of acetate would promote the precursor–substrate interactions, resulting in the formation of continuous, uniform, and adherent ZIF-8 TFs on copper foils. The resulting ZIF-8 TFs showed two orientations at the [110] and [211] directions and exhibited excellent mechanical and thermal stabilities. Sun et al. [102] fabricated high-quality MOF-TFs on α-Al_2O_3 ceramic tubes via a seed-assisted secondary growth strategy, in which the seeds were prepared via a solvent-vaporization driving force (Figure 3B). The ZIF-8 seeds were ready on the ceramic tube by pouring a stable precursor solution mixture of ZIF-8 into them and then sealing them inside the tube with a rubber stopper. The seeds then underwent heating at 55 °C for 4 h to react, at 55 °C for 1 h for drying, and at 25 °C for 12 h for washing. Then, the seeded ceramic tube was subjected to traditional solvothermal synthesis at 110 °C for 24 h to achieve a continuous ZIF-8 TF in the inner skin of the tube, which could be of interest for gas separation.

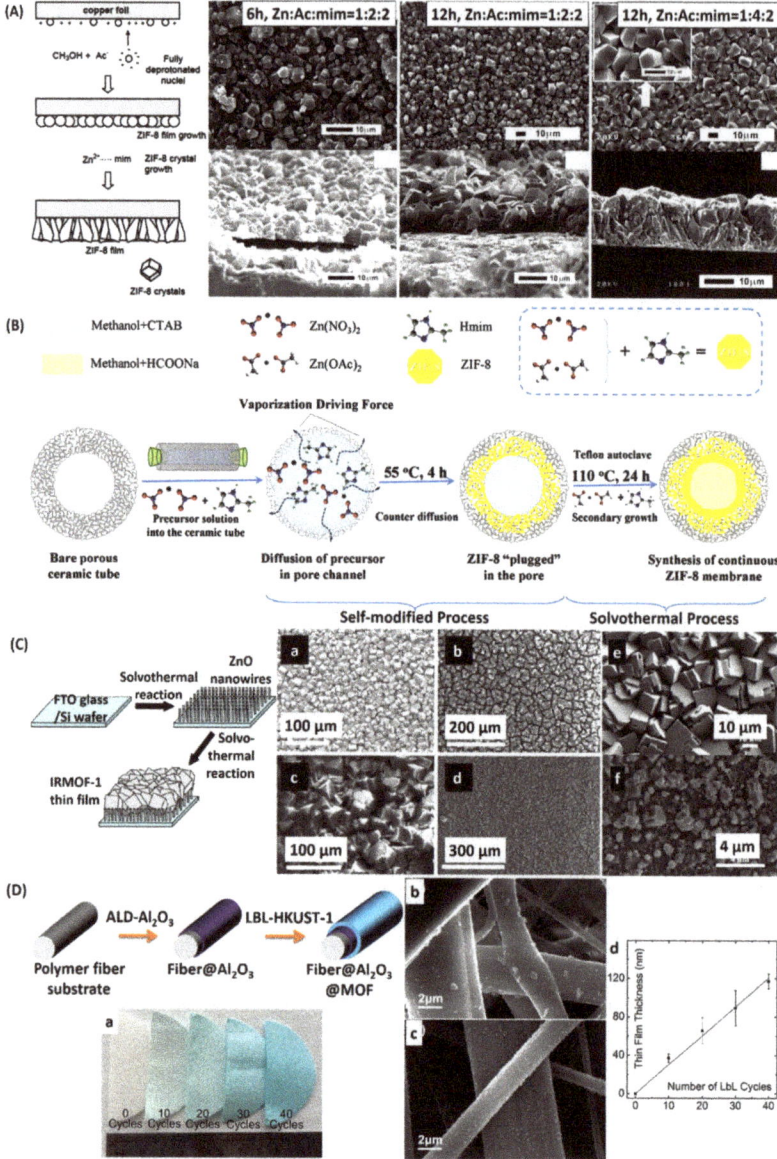

Figure 3. (**A**) Schematic of ZIF-8 growth on Cu substrates in methanol-based synthesis, and the top-view (**top**) and cross-section (**bottom**) SEM images of copper foils treated with different methanol-based

protocols. All scale bars, 10 μm. (Reproduced with permission from Papporello et al., Microporous and Mesoporous Materials; published by Elsevier BV, 2015.) (**B**) Preparation schematic of the counter-diffusion method for plugging pore and the secondary growth method for ZIF-8 film on the inner surface of a ceramic tube. (Reproduced with permission from Sun et al., RSC Advances; published by Royal Society of Chemistry, 2014.) (**C**) Schematic of the templated methodology of MOF-TF fabrication on ZnO NWs, and SEM images of (**a**) IRMOF-1, (**b**) IRMOF-3, (**c**) IRMOF-8, and (**d**) IRMOF-9 films grown on ZnO NWs, respectively, and the IRMOF-1 film by (**e**) traditional solvothermal synthesis and (**f**) microwave-assisted synthesis. (Reproduced with permission from Abdollahian et al., Crystal Growth & Design; published by American Chemical Society, 2014.) (**D**) Schematic of the LBL synthesis route. (**a**) Optical images of ALD-Al_2O_3-coated PP fibers with different LBL HKUST-1 TFs, and SEM images of an HKUST-1 TFs on (**b**) untreated and (**c**) ALD-Al_2O_3-coated PP fibers. (**d**) The thickness of the MOF-TFs on ALD-Al_2O_3-coated PP fibers measured from cross-section TEM images. (Reproduced with permission from Zhao et al., Journal of Materials Chemistry A; published by Royal Society of Chemistry, 2015.)

Another type of inorganic seeds is based on metal oxide nanostructures, such as NPs and nanowires (NWs). Abdollahian et al. [103] reported the fabrication of different IRMOF-TFs on ZnO NW-functionalized indium tin oxide (ITO) glass substrates, including glass, indium tin oxide (ITO) glass, and Si wafer (Figure 3C). ZnO NWs were grown on the substrate via a traditional solvothermal synthesis before immersion in each IRMOF precursor solution. The resulting IRMOF-TFs obtained high crystallinity after 20–24 h of secondary growth, showing a preferred out-of-plane orientation depending on the type of IRMOFs and acquiring μm-level thickness. All the IRMOF-TFs displayed an average thickness that was about 25 μm and exhibited similar morphology. In this study, the microwave-assisted synthesis showed its advantage in shortening the reaction time to 10 min for IRMOF-1 TF, which achieved about 1 μm in thickness. However, it was observed that the film morphology and crystallinity may be compromised by the rapid crystallization process as shown in the SEM images in Figure 3C,E,F. Furthermore, it was discovered that the metal oxide-based seeding layer does not necessarily contain the same metal species as that in the desired MOF-TFs. Zhao et al. [104] prepared Al_2O_3 seeding layers on polymer fibers via an ALD process to assist the nucleation and growth of copper-based HKUST-1 TFs (Figure 3D). Uniform HKUST-1 TFs were achieved on ALD-Al_2O_3-coated PP fibers via the LBL deposition strategy, by dipping the fibers in precursor solutions and solvent sequentially. A thorough rinsing by the solvent was necessary after each dipping in metal and organic precursors, respectively, to ensure complete removal of the unreacted precursors and unattached nuclei. Based on an HKUST-1 TF prepared by 40 LBL cycles in this study, the N_2 adsorption BET surface area could reach 535 m^2/g_{MOF} (93.6 $m^2/g_{MOF+fiber}$), and high dynamic loadings for NH_3 (1.37 $mol_{NH3}/kg_{MOF+fiber}$) and H_2S (1.49 $mol_{H2S}/kg_{MOF+fiber}$).

3.3. Layer-by-Layer Deposition

Although the traditional hydro/solvothermal synthesis is a classic synthesis strategy to obtain MOF-TFs, many obstacles remain in its development, such as the difficulty in controlling the fabrication process that may result in uncontrollable film thickness and discontinuous film formation, and the high cost from large reactant consumption and waste production that limits its application [81]. There are many existing synthetic strategies to address these challenges; the LBL deposition strategy shows excellent control in film thickness [105,106] and surface roughness [104].

In contrast to the traditional one-pot hydro/solvothermal synthesis, the solutions for metal and organic precursors are held separately in LBL deposition of MOF-TFs, and the substrate is placed in each precursor solution sequentially, leading the fabrication of MOF-TFs by depositing alternating layers of oppositely charged precursor species (i.e., the metal cations and the deprotonated organic linkers). The detailed synthesis conditions for different types of MOF-TFs may vary. The LBL deposition strategy is favorable for the fabrication of oriented and well-defined MOF-TFs, in particular for surface-mounted metal-organic frameworks (SURMOFs) [107]. Wang and Wöll [108] published an excellent review in

2019 about the fabrication methods of SURMOFs via programmed LBL assembly techniques, depicting the broad application of the LBL deposition strategy in the preparation of MOF-TFs.

Shekhah et al. [107] established a stepwise LBL deposition route for the fabrication of MOF-TFs (Figure 4A). Gold substrate was functionalized by MHDA, resulting in a COOH-terminated surface for the secondary growth of HKUST-1 TFs. The modified substrates were immersed in 1 mmol/L ethanolic solution of $Cu(CH_3COO)_2$ for 30 min, and 1 h in 1 mmol/L ethanolic solution of 1,3,5-benzenetricarboxylic acid, with a rinse between each immersion. Highly oriented HKUST-1 TFs showing the [100] growth direction were deposited on COOH-terminated SAMs on gold substrates. Shekhah et al. [109] also implemented LBL deposition in the fabrication of MOF-TFs on porous substrates. Crystalline and homogeneous HKUST-1 and ZIF-8 TFs were achieved on a confined surface of mesoporous SiO_2 foams, showing the potential of the LBL deposition strategy in controlling and directing the fabrication of MOF-TFs. Yao et al. [110] developed a spray-LBL deposition strategy to fabricate MOF-TFs (Figure 4B). The metal and organic precursor solutions for $Cu_3(HHTP)_2$, where HHTP = 2,3,6,7,10,11-hexahydroxytriphenylene, were sprayed on the substrate surface alternately to obtain the $Cu_3(HHTP)_2$ MOF-TFs. Like the traditional LBL strategy, the number of deposition layers can be facilely controlled by the spraying times. A good control over the film thickness was achieved, with a thickness increment of about 2 nm per spraying cycle. The performance in NH_3 room-temperature sensing was tested based on a 20-nm thick $Cu_3(HHTP)_2$ TF, which reached a limit of detection that was 100 ppm. Other than planar substrates, uniform MOF-TFs can also be fabricated on different types of substrates via the LBL deposition strategy [104], which also presents excellent control over the film thickness. As mentioned in the section of secondary growth, MOF-TFs designed on QCM substrates often rely on functional SAMs to assist fabrication via the LBL deposition strategy as it offers a simple fabrication process and high growth rates and realizes uniform and oriented MOF-TFs with controllable thicknesses and chemical compositions [107,111–113]. Stavila et al. [114] systematically studied the formation process of HKUST-1 TFs on QCM electrodes assisted by the LBL liquid-phase epitaxy (LPE) (Figure 4C). Wannapaiboon et al. [51] developed a direct fabrication strategy of MOF-TFs, with the hierarchical structure on the surface of QCM sensors, via an LBL-LPE process, which allows the adsorption performances of the heterostructures to probe in real time. Other than planar substrates, the fabrication of uniform and continuous MOF-TFs can also be made on arched surfaces, such as optical fibers [42], via the LBL deposition manner.

Overall, the LBL deposition strategy can realize the fabrication of many types of MOF-TFs and enable precise control of the amount and location/distribution of functionalities for tailored properties, presenting a huge potential to extend the already considerable flexibility of MOFs. This synthesis strategy also provides a pathway to better understand the heterogeneous nucleation and growth of MOF-TFs [57], offers good control of the film thickness, gives rise to highly oriented and uniform TFs, and enables the fabrication of more complex MOF heterostructures [115–117]. However, this type of synthesis strategy is limited to the fabrication of specific types of MOF-TFs and is only available on solid supports [118], which restricts its application.

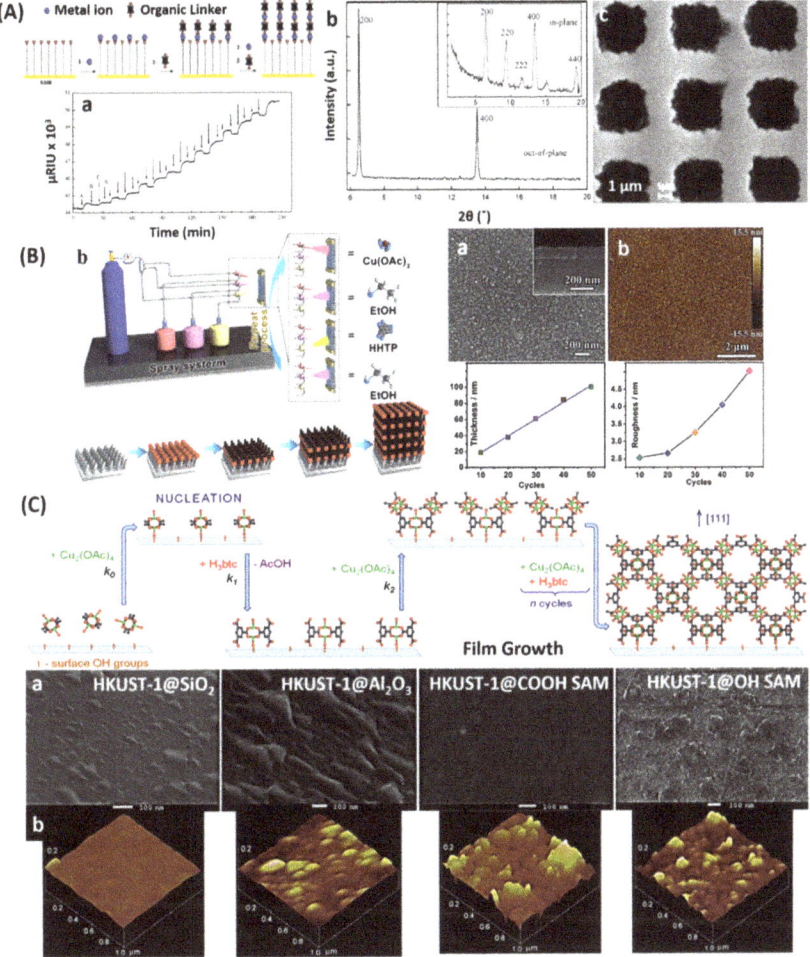

Figure 4. (**A**) Schematic of the step-by-step growth of HKUST-1 TF on a COOH-terminated SAM, and (**a**) the corresponding SPR signal as a function of time recorded in situ during sequential injections of Cu(OAc)$_2$, ethanol, and 1,3,5-benzenetricarboxylic acid. (**b**) XRD data of an HKUST-1 TF (40 cycles) grown on COOH-terminated SAM, inserted the in-plane data. (**c**) SEM image of HKUST-1 (40 cycles) grown on an SAM laterally patterned by μCP consisting of COOH-terminated squares and CH$_3$-terminated stripes. (Reproduced with permission from Shekhah et al., Journal of the American Chemical Society; published by American Chemical Society, 2007.) (**B**) A schematic diagram to illustrate the preparation of Cu$_3$(HHTP)$_2$ TF via spraying. (**a**) SEM and (**b**) AFM images of a Cu$_3$(HHTP)$_2$ TF, and the corresponding film thickness and surface roughness. (Reproduced with permission from Yao et al., Angewandte Chemie International Edition; published by Wiley-VCH, 2017.) (**C**) Schematic of the proposed model for HKUST-1 nucleation and growth on oxide surfaces (Cu-green, O-red, C-gray). (**a**) SEM and (**b**) AFM images of HKUST-1 TFs (40 cycles) on different substrate surfaces. (Reproduced with permission from Stavila et al., Chemical Science; published by Royal Society of Chemistry, 2012.)

3.4. Dip-Coating Deposition

Dip coating (DC) is a simple, low-cost, and reproducible method for fabricating TFs and is extensively used in industries. This method is also applicable in the fabrication of MOF-TFs based on a colloidal suspension of MOFs NCs [119,120]. In this synthesis strategy, the metal and organic precursor solutions are well mixed to form a uniform colloidal suspension, in which a substrate is put in place to obtain an MOF-TF. Subjected to the continuous growth of MOF crystals in the suspension, the precursor solution mixture needs to be replaced after a certain time to ensure a high enough concentration of reactants for film growth. The reaction time for different MOFs varies depending on the reaction conditions. The DC technique can also be implemented in an LBL deposition manner, where the thickness of MOF-TFs can be controlled by modifying the immersing time, the number of dipping cycles, and the withdrawing speed [121].

Horcajada et al. [122] developed a colloidal route for the fabrication of MOF-TFs on Si wafers via DC deposition. In the fabrication of MIL-89 ($Fe_3O(CH_3OH)_3[O_2C-(CH)_4-CO_2]_3Cl(CH_3OH)_6$) MOF-TFs, a colloidal solution iron(III) acetate and muconic acid in ethanol was prepared first by heating up the mixture solution at 60 °C for 10 min to promote the formation of colloidal particles. Each deposition cycle contained 2 min for immersion before the withdrawal with a speed of 4 mm/s at 15% relative humidity. After each cycle, the TF was washed with ethanol and dried either at room temperature or at 130 °C in air for 5 min. Flexible and uniform MIL-89 MOF-TFs were obtained (Figure 5A) under a consistent growth rate of 40 nm/coating. Then, Lu and Hupp [123] implemented DC deposition in the fabrication of ZIF-8 TFs. Highly oriented and continuous ZIF-8 TFs were achieved directly on glass slides and Si wafers (Figure 5B). The controllable synthesis showed a linear growth rate of ZIF-8 TFs of 100 nm/coating (30 min/cycle) and presented a thickness-dependent color-changing property that was interesting in optics. The DC deposition strategy is used extensively to fabricate various MOF-TFs that are of interest in electronics and optics [16,43,121–124].

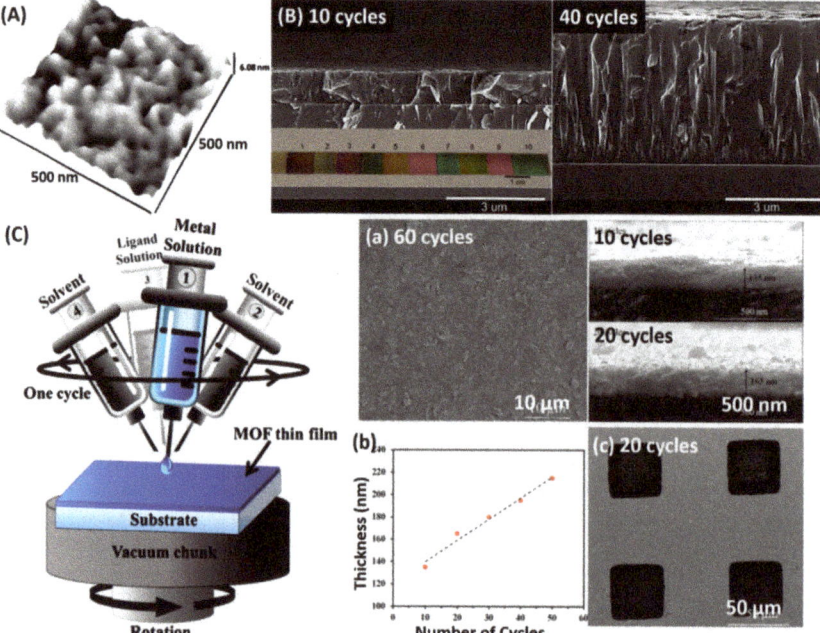

Figure 5. (**A**) AFM image of a MIL-89(gel) TF *via* the DC method. (Reproduced with permission from Horcajada et al., Advanced Materials; published by Wiley-VCH, 2009.) (**B**) SEM images of ZIF-8 films grown on Si substrates with different cycles of dip coating, inserted is the photograph of a series of ZIF-8 films of various thicknesses grown on Si substrates. (Reproduced with permission from Lu et al.,

Journal of the American Chemical Society; published by American Chemical Society, 2010.) (**C**) Schematic for the fabrication of MOF-TFs using the LPE approach adapted to the SC method. (**a**) Top-view and cross-section SEM images of HKUST-1 TFs with different deposition cycles. (**b**) Height profile of the HKUST-1 TF from different deposition cycles. (**c**) SEM image of HKUST-1 TF grown by 20 cycles on a SAM laterally patterned by µCP consisting of COOH-terminated squares and CH$_3$-terminated stripes. (Reproduced with permission from Chernikova et al., ACS applied materials & interfaces; published by American Chemical Society, 2016.)

3.5. Spin-Coating Deposition

Spin coating (SC) is a commonly used method to apply a uniform TF onto a flat solid substrate, which is also applicable for the fabrication of MOF-TFs. In a typical process of making MOF-TFs via the SC method, different precursor solutions for MOFs are dropped on the center of a flat substrate on a spinning object, which is set to a certain spinning speed and time, and the volume/concentration of the solutions are controlled to achieve a uniform distribution of precursor on the substrate surface [121].

Chernikova et al. [125] proposed an LBL assembly strategy via an SC process, which realized the fabrication of smooth MOF-TFs in a relatively short time (Figure 5C). In this fabrication process, a spin coater was equipped with four micro-syringes containing the precursor solutions and solvents separately. First, the metal precursor solution was applied to the spinning substrate by one syringe. After it was uniformly distributed, the solvent was applied using another syringe to rinse the substrate surface. These two steps were repeated to achieve the coating of the organic precursor solution. The spinning time was 5 s for each solution and 8–10 s for each solvent depending on the type of MOFs. For each step, only 50 µL of liquid was used. The four steps finished one cycle of SC deposition. One can increase the thickness of MOF-TFs through multiple cycles. Different MOF-TFs, such as HKUST-1, ZIF-8, Cu$_2$(BDC)$_2$, and Zn$_2$(BDC)$_2$, were achieved via SC deposition. In this study, a Cu$_2$(BDC)$_2$ TF obtained a thickness of about 140 nm for 10 cycles. Moreover, it was found that the time needed to finish 100 cycles was only 50 min, which was significantly shorter in comparison to the 25 h needed in the conventional LBL deposition process. This strategy is extendable to the fabrication of other MOF structures, such as hybrid MMMs [126].

The SC technique can achieve MOF-TFs in a short time with low reactant consumption, and the resulting MOF-TFs can be dense and uniform, with thicknesses that can range from the micron to nano-scale [125]. However, it is appreciable that this method could cause crystal defects, and results in structural defects in MOF-TFs [121].

3.6. Interfacial Synthesis

The interface synthesis of MOF-TFs can be realized at the interface between two immiscible media, such as oil and water or air and water, which could result in the formation of freestanding MOF-TFs.

The interfacial synthesis could occur at a liquid–liquid interface, where the metal and organic precursors are dissolved in two immiscible solvents separately. Ameloot et al. [80] reported the first demonstration of liquid–liquid interfacial synthesis for MOF-TFs. The coordination of Cu^{2+} cations and BTC^{3-} linkers happened at the interface of an aqueous solution containing the metal precursor and an organic solution containing the organic precursor, thus resulting in a freestanding and uniform HKUST-1 TF at the interface. Other types of MOF-TFs, such as ZIF-8 TFs [127], can also be obtained through this type of synthesis strategy.

Other than the synthesis at the interface of two immiscible liquids, there is an air–liquid interfacial synthesis for MOF-TFs. Li et al. [128] developed a fabrication strategy of MOF-TFs, patterns, and layered structures (including hybrid layers) via a templated air–liquid interfacial synthesis strategy (Figure 6). The resulting ZIF-8 TFs obtained hierarchical structures through close-packed arrays of colloidal spheres floating at the air–solution interface, which demonstrated different physical properties from unstructured TFs. The obtained 2D-ordered macroporous (2DOM) ZIF-8 TF-loaded polyvinylidene fluoride (PVDF) membrane showed a much improved experimental separation factor,

3.33, for methyl blue and methyl orange in comparison with pristine PVDF (1.64) and unstructured ZIF-8 TF-loaded PVDF (3.01). Other than adsorption and separation, this well-organized superstructure was also of interest in catalysis and microreactors. This method introduced a facile fabrication of multicomponent devices and could readily extend to other types of templates and MOFs.

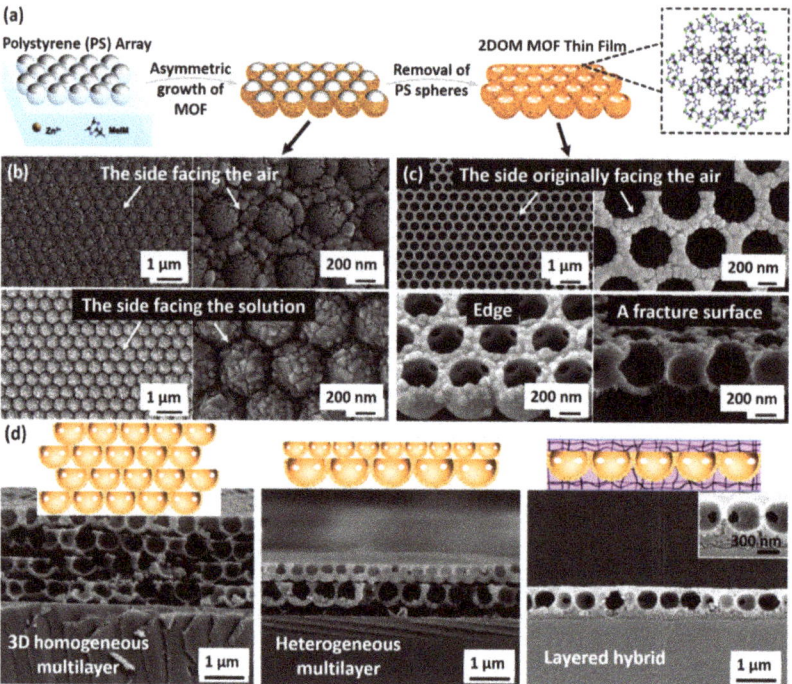

Figure 6. (a) Schematic of the asymmetric growth of MOF-TFs on 2D arrays anchored at the air–liquid interface and the fabrication of 2DOM MOF-TFs. SEM images of (b) ZIF-8 formed at the air–liquid interface of different sides, and (c) ZIF-8 TFs obtained after removal of polystyrene spheres. (d) Schematic and corresponding SEM images of vertically layered architectures based on transferable MOF superstructures. (Reproduced with permission from Li et al., Crystal Growth & Design; published by American Chemical Society, 2016.)

3.7. Contra-Diffusion Synthesis

Although many strategies have been developed to achieve a high rate of heterogeneous nucleation of MOFs on substrates, they often complicate the process [60,129,130]. Therefore, a contra-diffusion synthesis strategy was developed to realize an efficient way for the fabrication of MOF-TFs. The contra-diffusion synthesis is similar to the interfacial synthesis of MOF-TFs in that the metal and organic precursor solutions are placed separately. However, a contra-diffusion synthesis requires porous substrates [131–135] whereas the interfacial synthesis could proceed without one [127]. In the contra-diffusion synthesis, the substrate separates the two precursor solutions. The precursors diffuse in opposite directions through the substrate, and MOFs can form upon encountering the precursors. The contra-diffusion synthesis is capable of making MOF-TFs that are embedded into the porous substrates, resulting in a strong adhesion between MOF-TFs and substrates. Although it exhibits simplicity and good reproducibility, the contra-diffusion synthesis strategy is limited to MOF-TFs with high permeability and is only achievable on porous substrates, which limits its application.

Yao et al. [133] developed a synthesis strategy to fabricate ZIF-8 TFs on flexible substrates via a contra-diffusion strategy (Figure 7A). Two precursor solutions containing Zn^{2+} and 2-mIm, respectively, diffused from opposite sides of porous nylon substrate in opposite directions, and ZIF-8 TFs were formed on both sides of the substrate. In this type of synthesis, ZIF-8 TFs revealed different morphologies on two sides of the nylon substrate, which could arise from different local molar ratios of 2-mIm/Zn^{2+}, where large crystals with sizes of 0.2–5 µm formed on the Zn^{2+} side and small NCs formed on the 2-mIm side as shown by the SEM images in Figure 7A. After crystallization at room temperature for 72 h, the surficial ZIF-8 film could reach 16 µm at the Zn^{2+} side, which exhibited a H_2/N_2 ideal selectivity of 4.3 with H_2 permeance of 1.97×10^{-6} mol/m^2sPa and N_2 permeance of 0.46×10^{-6} mol/m^2sPa. One can control the coverage of the porous substrate and the thickness of MOF-TFs by tuning the reaction time and the metal/linker concentration ratio in the reaction system [92].

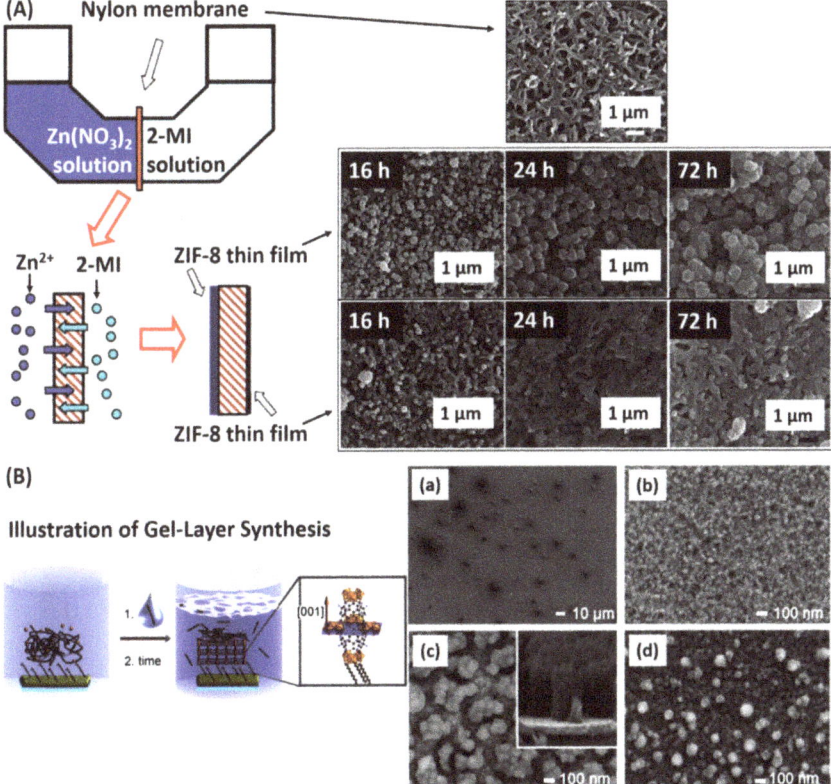

Figure 7. (**A**) Diffusion cell for ZIF-8 film preparation and the schematic formation of ZIF-8 films on both sides of the nylon support via the contra-diffusion method, and corresponding SEM images of a bare nylon membrane and ZIF-8 TFs formed on different sides of nylon membranes at room temperature. (Reproduced with permission from Yao et al., Chemical communications; published by Royal Society of Chemistry, 2011.) (**B**) Illustration of the gel-layer method to fabricate an oriented metal–organic framework TF on an SAM-functionalized Au substrate. SEM images of (**a**) HKUST-1 on an OH-terminated substrate and (**b**) Fe-MIL-88B-NH_2 TF on a COOH-terminated substrate. (**c**) A thick Fe-MIL-88B-NH_2 film with island formation under a higher iron(III) concentration, and (**d**) larger single crystals formed on the film surface with higher molecular weight poly(ethylene oxide) (10^5). (Reproduced with permission from Schoedel et al., Angewandte Chemie International Edition; published by Wiley-VCH, 2010.)

3.8. Gel-Layer Synthesis

Schoedel et al. [136] developed a gel-layer strategy for the fabrication of MOF-TFs (Figure 7B). In this method, thin poly(ethylene oxide) or poly(ethylene glycol) gel layers hold a high concentration of metal precursors near SAM-coated gold substrates. Following the diffusion of organic linkers through the metal-containing gel, heterogeneous nucleation of MOFs occurs at the gel–SAM interface. Both rigid HKUST-1 TFs and flexible Fe-MIL-88B-NH_2 TFs were successfully fabricated on SAM-coated gold substrates after 96 h of reaction at room temperature. The resulting Fe-MIL-88B-NH_2 TFs were homogeneous and showed a preferred orientation at the [001] direction based on the COOH-terminated SAMs; however, the HKUST-1 TFs did not. As shown by the SEM images in Figure 7B, the resulting thickness of Fe-MIL-88B-NH_2 TFs was 500–550 nm, consisting of small islands of crystals with several hundred nm in diameter, of which the gaps between islands were about 100 nm. There was also a very homogeneous layer at the bottom, of which the thickness was about 40 nm.

The gel-layer synthesis method can realize the conservation of a high concentration of reactants to be employed in heterogeneous film formation and eliminates the necessary precondition of traditional hydro/solvothermal precursor solutions. The molecular weight of the polymer (gel) and the concentration of metal precursors in the gels are the keys to control the morphology and thickness of the resulting MOF-TFs. This method is pictured to be widely applicable with suitable gel matrices, yet it could be very time-consuming, depending on the types of MOFs and gel matrices.

3.9. Evaporation Method

The evaporation method is an effective way to synthesize MOF-TFs with a controllable scale via an evaporation-induced crystallization process [137,138]. In this synthesis strategy, a clear and stable precursor solution is prepared without small MOF nuclei. Following the removal of the solvent by slow evaporation, there is crystallization that leads to the formation of MOF crystals locally on the substrate. This method can achieve a precise localization of MOF crystals on solid substrates.

Ameloot et al. [137] reported the fabrication of HKUST-1, MOF-5, and ZIF-8 patterned TFs via the evaporation method in a stamping manner (Figure 8A). The precursor solution of each type of MOFs was prepared as in conventional solvothermal synthesis. Patterning was performed by placing stamps inked with the precursor solution on a glass substrate. In this method, the stamps were removed after solvent evaporation. Unlike conventional secondary growth based on SAMs, the resulting HKUST-1 crystals in this study showed [111] growth orientation regardless of the substrate surface termination (i.e., silanol, vinyl, and carboxylic acid groups). It was suggested by the authors that the confinement between the stamp and the substrate during in situ crystallization had a more significant influence on the preferred orientation of the resulting MOF-TFs than the substrate surface chemistry. It was observed that the roughness of all films obtained via the evaporation method appeared high. Zhuang et al. [138] reported the fabrication of HKUST-1 TFs on rigid substrates at room temperature via an evaporation process (Figure 8B), and then they implemented this method with an inkjet-printing technique and achieved patterned MOFs on flexible substrates as shown in Figure 8C [139]. The evaporation method could also be applied with the drop-casting method to make MOF-TFs [140]. Other types of MOF-TFs, such as ZIF-7 $(Zn(bIm)_2)$ [141], are also achievable via fast evaporation of the solvent.

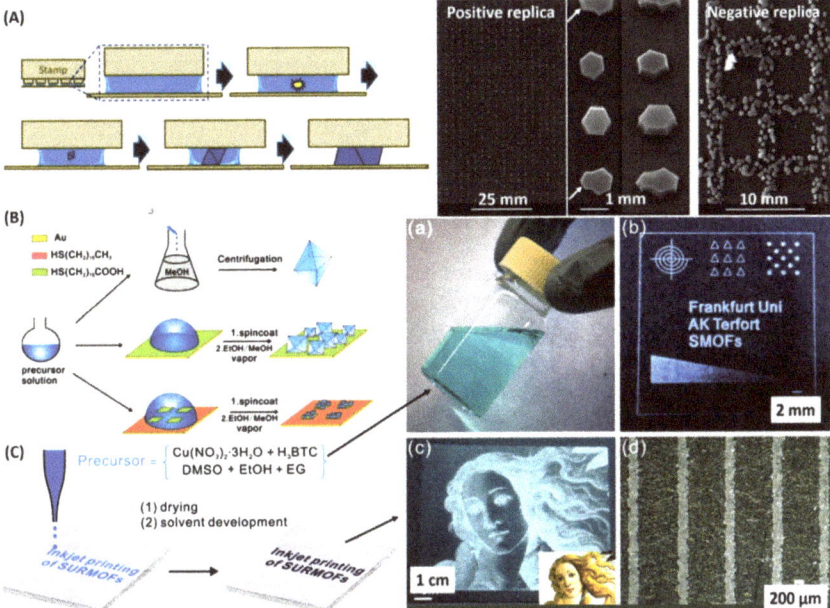

Figure 8. (**A**) Schematics of the nucleation, growth, and orientation of HKUST-1 crystals in confinement during solvent evaporation, and SEM images of the patterned deposition of HKUST-1 from the positive and negative replica. Arrows indicate intergrowths caused by the second nucleation. (Reproduced with permission from Ameloot et al., Advanced Materials; published by Wiley-VCH, 2010.) (**B**) Schematics of the synthesis of bulk HKUST-1 crystals and SC fabrication of highly oriented TFs and patterns. (Reproduced with permission from Zhuang et al., Advanced Functional Materials; published by Wiley-VCH, 2011.) (**C**) Schematics for inkjet-printing SURMOFs onto flexible substrates using an HKUST-1 precursor solution as "ink". Optical photos of (**a**) an HKUST-1 ink solution, (**b**) various patterns, letters, and a gradient wedge printed onto polyethylene terephthalate foil, (**c**) Botticelli's "Venus," which was printed in HKUST-1 (the inset shows the original image), and (**d**) a line array (2 cycles). (Reproduced with permission from Zhuang et al., Advanced Materials; published by Wiley-VCH, 2013.)

These contributions demonstrate the potential of the evaporation method in placing MOFs in various microprinting and nanotechnological fields and that faster solvent evaporation could produce more defective and less stable films compared to the controlled release of solvent in a traditional process [142], whereas the films could potentially enhance gas permeation [126].

4. General Liquid–Solid Synthesis

Although the liquid–liquid synthesis strategies have been extensively studied and widely used in the fabrication of MOF-TFs, the limited scalability of these solution-based approaches prevent their large-scale production. Many other synthesis strategies were developed to realize a more economical and greener process. The liquid–solid reaction, one of the alternatives, can significantly promote heterogeneous nucleation and growth of MOF-TFs on substrates.

4.1. Electrochemical Deposition

Electrochemical deposition (ECD) is one emerging synthesis strategy conducted in the fabrication of MOF-TFs, as discussed in many excellent reviews [72,118,143]. In a classic ECD system for MOF-TF deposition, there is a two-electrode cell (a three-electrode cell is also available) containing metals, organic linkers, and electrolytes, and the MOF-TF is formed on the electrode(s) via the coordination of metal ions with deprotonated organic ligands near the electrode surface through anodic dissolution [144–146], cathode reduction [147,148], or charge driving [149,150]. This type of synthesis strategy allows for the fabrication of MOF-TFs with controllable thicknesses via real-time monitoring of the amount of passed charge. Moreover, the electrochemical nature of this process offers an in situ repairing of defects, such as cracks and pinholes. However, this method is limited to the fabrication of non-conductive MOF-TFs on conductive substrates, which restricts its application. Besides, metal ions with high inertness could separate on the cathode, and the organic linkers may be oxidized.

Anodic electrodeposition (AED) that is based on the dissolution of metal components is the method of choice for large-scale production of some commercially available MOFs because of its scalability, ease of processing, and low working temperature. Joaristi et al. [151] reported the fabrication of several archetypical MOF-TFs via anodic dissolution in an electrochemical cell (Figure 9A). The anode was a metal plate (highly pure Zn, Cu, or Al), and the cathode was preferable inter alia (i.e., Zn-Zn, Cu-Cu, and Al-Al), while graphite and steel were also available. MOF-TFs, including HKUST-1, ZIF-8, MIL-100(Al) ($Al_3O(H_2O)_2(BTC)_2X$, X = OH or F), and MIL-53(Al) ($Al(OH)[O_2C-C_6H_4-CO_2]$), were obtained via AED in an electrodeposition cell containing proper organic linkers. Take the fabrication of HKUST-1 TFs as an example. A linker solution was prepared by dissolving BTC and tributylmethylammonium methyl sulfate in 96 vol % ethanol. The solution was heated up to 80 °C in the electrochemical cell with two copper electrodes spaced at least 3 cm apart. Then, 50 mA was passed through the system for 1 h. The product was filtered off and cleaned with ethanol at room temperature overnight, then filtered again and dried at 100 °C. An average of ~100 mg of dried HKUST-1 was obtained from each synthesis. As shown in Figure 9Ac, HKUST-1 TFs covered the entire copper mesh. The deposition conditions, such as linker solubility, temperature, and current density, could be adjusted to obtain the best morphology, coverage, and crystallinity for the resulting MOF-TFs. Compared to traditional hydro/solvothermal synthesis, ECD-based synthesis strategies showed advantages in reducing the reaction time and temperature in this study. AED is also accessible in the fabrication of MOF-TFs on other types of conductive substrates. Hauser et al. [152] fabricate MOF-TFs on ITO glass. First, copper or zinc microcrystalline films (Cu-ITO and Zn-ITO) were deposited electrochemically on ITO electrodes in aqueous/ethanol salt solutions serving as the metal source in anodic dissolution. Several well-adhered and homogeneous MOF-TFs, including HKUST-1, $Cu(C_{10}H_8N_2)Br_2$, Zn-BTC (Zn_3BTC_2), and Zn_2BPDC (where BPDC = 2,2'-bipyridine-5,5'-dicarboxylate), were obtained on ITO anodes under the deposition conditions in this study. HKUST-1 TFs can be achieved on copper-coated QCMs via the AED strategy as well [144]. While most MOF-TFs can be deposited using metal or a metal-coated electrode, it is difficult and expensive to electrochemically deposit rare earth metals as the electrodes for the fabrication of luminescent MOF (LMOF)-TFs. Li et al. [145] developed a microwave-assisted electrochemical deposition strategy for LMOF-TFs (Figure 9B). In this study, a dense and homogeneous $Ln(OH)_3$ (Ln = Eu, and Tb) layer was first deposited on a fluorine-doped tin oxide (FTO) electrode after eight cycles of ECD in a cell containing $Ln(NO_3)_3$. Subsequently, the $Ln(OH)_3$ layer was converted to Ln-MOF TFs under microwave irradiation in a cell containing TPO linkers, where TPO = tris-4-carboxylatephenyl phosphineoxide. Patterned Ln-MOF TFs were also made using this microwave-assisted ECD strategy by patterning poly(dimethylsiloxane) (PDMS) films on FTO glass. These patterned TFs had strong luminescence properties, which is of interest in the fields of color displays, luminescent sensors, anti-counterfeiting barcodes, and structural probes. In summary, in the AED strategy for MOF-TFs, the electrode acts as the cation source, and there are blocking-free pores and easily controlled metal oxidation states. However, corrosion of the anode is inevitable during the AED process, and there are restrictions on the anode material selection and single-phase MOF conformation.

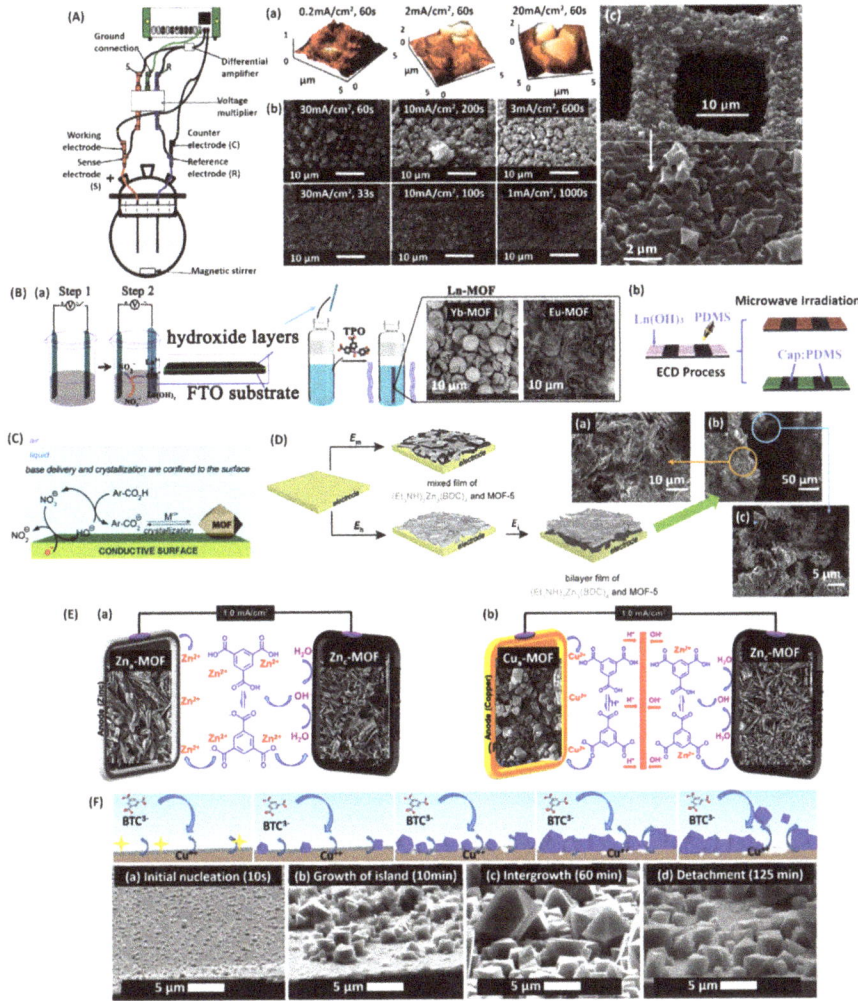

Figure 9. (**A**) Schematic view of an electrochemical synthesis cell. (**a**) AFM and (**b**) SEM images of HKUST-1 TFs fabricated on copper electrodes and (**c**) on a copper mesh under different electrochemical conditions. (Reproduced with permission from Martinez et al., Crystal Growth & Design; published by American Chemical Society, 2012.) (**B**) Schematic illustration of (**a**) ECD for $Ln(OH)_3$ layers on a transparent FTO glass and the microwave conversion to Ln-MOFs, and (**b**) the patterning growth of luminescent barcodes. (Reproduced with permission from Li et al., Chemical communications; published by Royal Society of Chemistry, 2016.) (**C**) Mechanism of CED. (Reproduced with permission from Li et al., Journal of the American Chemical Society; published by American Chemical Society, 2011.) (**D**) Schematic illustration of the formation of a biphasic mixed film at (cathodic) potential, E_i. $E_i < E_m < E_h$. (**a–c**) SEM images of (**b**) a film produced by sequential growth at 1.10 and 1.50 V, displaying (**a**) the characteristic feather-like morphology of $(Et_3NH)_2Zn_3(BDC)_4$ in the top layer and (**c**) the small crystallites associated with the Zn/MOF-5 composite in the layer closer to the electrode surface. (Reproduced with permission from Li et al., Chemical Science; published by Royal Society of Chemistry, 2014.) (**E**) (**a**) Zn_a/Zn_c-MOF-TFs modified electrodes by the CPED and (**b**) Cu_a/Zn_c-MOF-TF

modified electrodes by the DPED at I_{app} = 1 mA/cm^2, t = 10,800 s; inserted the corresponding SEM images of the modified electrodes. (Reproduced with permission from Alizadeh et al., Scientific reports; published by Nature Research, 2019.) (**F**) Proposed mechanism of AED and (**a–d**) SEM images of the four phases. (Reproduced with permission from Campagnol et al., Journal of Materials Chemistry A; published by Royal Society of Chemistry, 2016.)

Cathodic generated –OH groups can promote the deprotonation of organic linkers in the typical cathodic electrodeposition (CED) process. At the same time, metal ions move to the surface of the cathode and lead to the formation of MOF-TFs locally on the cathode. Li et al. [147] established the CED for the fabrication of crystalline MOF-TFs directly on conductive surfaces (Figure 9C). In this study, a substantial amount of –OH groups were generated and accumulated near the surface of the FTO electrode, which promoted the nucleation and growth of MOF-5 TFs exclusively onto the conductive FTO glass. Then, they demonstrated the fabrication of multiphasic and multilayered MOF-TFs via the same synthesis strategy (Figure 9D) [148]. In this CED experiment, the electrolyte solution was prepared by dissolving tetrabutylammonium hexafluorophosphate in DMF and stored in a sealed bottle in a nitrogen-filled glovebox. A typical deposition solution in this study consisted of 100 mmol/L Et$_3$NHCl, 100 mmol/L Zn(NO$_3$)$_2$, and 50 mmol/L H$_2$BDC. A bilayer film consisting of two different types of MOFs was obtained; this method offers a potential path to large-scale fabrication of heterostructured, multiphasic, and multilayered MOF-TFs. In contrast to the AED, the CED allows for the free choice of electrode material, in situ deprotonations of organic linkers, and multiphasic MOF-TF fabrication. However, this method is limited by possible metal reduction and pore blocking.

Paired electrodeposition (PED), which involves the pairing of both AED and CED strategies in the fabrication of MOF-TFs on both electrodes, is a newly developed strategy that realizes green synthesis of MOF-TFs based on the traditional ECD. Alizadeh et al. [153] developed convergent and divergent PEDs (CPED and DPED, respectively) for the simultaneous fabrication of MOF-TFs on both electrodes (Figure 9E). In the CPED, Zn-BTC MOF-TFs were obtained on both the zinc anode and carbon cathode. In the DPED, since they share the same type of organic linker, Zn-BTC and HKUST-1 TFs were obtained on the graphite cathode and copper anode, respectively. Take the fabrication of Zn$_a$/Zn$_c$-MOF-TFs as an example. To achieve CPED, both metal salt and metal were used as two cation sources, in which Zn(NO$_3$)$_2$ for CED and Zn metal as a sacrificial anode to generate Zn^{2+} cations for AED. Zn(NO$_3$)$_2$ as a cation source and NaNO$_3$ as a supporting electrolyte were dissolved in water (solution A, pH 2.1), and H$_3$BTC in ethanol (solution B). The prepared solutions were aged under stirring for 2.5 h at room temperature. Then, the electrodeposition process was performed in a homemade undivided two-electrode cell with Zn plate and carbon plate as the electrodes. After applying 1 mA/cm^2 for 3 h, complete coverage of Zn$_a$/Zn$_c$-MOF-TFs on the two electrodes was achieved, as shown in Figure 9Ea. This deposition strategy realized a current efficiency that was twice as much as the traditional methods, demonstrating a sustainable development for the fabrication of MOF-TFs using ECD.

Campagnol et al. [154] systematically investigated the fabrication mechanisms of MOF-TFs using ECD, including both AED and CED (Figure 9F). The proposed mechanism involved four phases: (I) Initial nucleation, (II) growth of MOF islands, (III) intergrowth, and (IV) crystal detachment. Hence, based on the understanding of the ECD mechanism, the metal used for the SBU needs to be carefully considered. For example, one could consider if the metal is noble or not, if it passivates or makes hydroxides. When these conditions are clear, the choice of AED or CED becomes easier. MOF-TFs (e.g., HKUST-1) can be made either way, although the resulting morphology can vary remarkably. Overall, the ECD strategies provide a roadmap for large-scale fabrication of MOF-TFs, including single-phase and heterostructured multiphasic and multilayered MOF-TFs and membranes.

4.2. Self-Sacrificing Templated Synthesis

MOF-TFs could be fabricated based on self-sacrificing solid templates consisting of metals, metal oxides, or hydroxides. In a general self-sacrificing templated synthesis, the template containing metal species of the desired MOF provides the metal cations for the formation of MOF-TFs, and meanwhile serves as the substrate, while the reaction solution provides only the organic linkers. In such a synthesis system, the formation of free MOF crystals in the liquid phase can be eliminated, and the fabrication of MOF-TFs on the template can be significantly promoted under optimal reaction conditions [155]. However, this reaction is limited by the dissolution of metal ions and may result in self-termination of the reaction when the MOF-TFs formed at the interface of the metal template and organic precursor solution becomes thick.

One type of self-sacrificing template consists of only metals. Zou et al. [156] reported a self-sacrificing templated fabrication for MOF-TFs using a single metal source. In this study, a Zn wafer was used as the template, which was first activated by H_2O_2 to form a hydroxide layer on the surface. The prepared Zn wafer was placed at the bottom of an autoclave filled with aqueous H_3BTC solution. Zn-BTC MOF-TFs were obtained on top of the remaining Zn wafer via hydrothermal synthesis at 140 °C for 24 h. After being cleaned with distilled water and dried at 85 °C, the obtained TF was tested for chemical sensing based on the photoluminescence properties of Zn-BTC MOFs. In this study, the Zn-BTC MOF-TFs showed high sensitivity and selectivity towards dimethylamine. A limit of detection for dimethylamine of 8.57 ppm was identified. Other types of MOF-TFs, such as HKUST-1, ZIF-8, Cu-BDC, and MOF-5 [157] (Figure 10A), have also been achieved on proper metal templates via this synthesis strategy upon substrate surface activation. Kang et al. [158] developed a self-sacrificing templated fabrication using a single metal source without activation pretreatment. A nickel net was used as the nickel source and substrate to synthesize Ni-MOF-TFs. Highly crystalline and continuous $Ni_2(L-asp)_2(bipy)$ (L-asp = L-aspartic acid; bipy = 4,40-bipyridine) MOF-TFs were then obtained locally on the nickel net via a traditional solvothermal synthesis at 150 °C after 48 h. The obtained mesoporous and microporous structure showed good separation results of (R)-2-methyl-2,4-pentanediol and (S)-2-methyl-2,4-pentanediol at high temperatures. When the operating temperature increased from 25 to 200 °C under 0.1 MPa, the permeance of R increased from 526 to 1047 g/m^2h, while that of S only slightly increased from 406 to 533 g/m^2h.

Another type of self-sacrificing template relies on metal oxides or hydroxides. Zhan et al. [159] reported the fabrication of ZIF-8 TFs based on ZnO templates (Figure 10B). By controlling the reaction conditions, ZnO@ZIF-8 core-shell nanorod (NR)/nanotube (NT)-structured TFs were obtained via traditional solvothermal synthesis at 70 °C for 24 h, where ZnO NRs/NTs served as the source of Zn^{2+} ions and as the template for ZIF-8 TFs. It was discovered that the solvent composition and reaction temperature are critical in this type of synthesis strategy since they can influence the dissolution rate of ZnO and the coordination rate of 2-mIm with the released Zn^{2+} ions. The ZnO@ZIF-8 core-shell structure showed different photocurrent responses for hole scavengers with various sizes, which was tested on H_2O_2 and ascorbic acid (AA). In the study, AA did not produce a similar enhancement effect to the photocurrent response like H_2O_2 because its molecule size is larger than the pore aperture of ZIF-8. Khaletskaya et al. [160] performed the fabrication of ZIF-8 TFs on QCMs via self-sacrificing synthesis. ZnO TFs were first prepared on the surface of QCM via ALD or magnetron sputtering, which resulted in different film morphologies for ZIF-8. The ZIF-8 TFs formed on sputtered ZnO templates showed no characteristic shape of ZIF-8 crystals and appeared larger than the one on the ALD-ZnO template, which was possibly due to the larger size of ZnO particles obtained by the sputtering process. The ZIF-8 TFs formed on ALD-ZnO templates showed a typical faceted shape of ZIF-8 crystals, which could result from the less dense and predominantly nanocrystalline ZnO precursor via ALD that provided a faster dissolution and conversion rate of the metal oxide. With either ZnO template, compact and homogeneous ZIF-8 TFs were obtained on both Si wafers and QCMs via microwave-assisted solvothermal synthesis at 80 °C for 1 h. In this study, highly oriented Al(OH)(NDC) MOF-TFs, where NDC = 1,4-naphthalenedicarboxylate, was achieved by using the same synthesis strategy. Mao et al. [161] reported the fabrication of HKUST-1 TFs on porous anodic aluminum oxide (AAO) substrates via a secondary growth based on $Cu(OH)_2$ nanostrands (CHNs). A highly positively charged CHN TF was prepared by mixing the 2-aminoethanol solution with $Cu(NO_3)_2$ solution and aging at room temperature, followed by filtering onto the surface of porous AAO. Then, the prepared CHN@AAO substrate was immersed in the H$_3$BTC solution for 1 h to obtain a thin layer of HKUST-1 crystals as the seeding layer to achieve HKUST-1 TF via secondary growth. After 24 h of reaction at 120 °C, dense HKUST-1 TF with a thickness of about 30 µm was obtained on the AAO substrates. This structure showed a high permeance of H$_2$ at room temperature of 3.14×10^{-6} mol/m^{2}sPa, which was higher than CH$_4$, N$_2$, O$_2$, and CO$_2$. It provided separation factors of the binary gases, i.e., H$_2$/CH$_4$, H$_2$/N$_2$, and H$_2$/CO$_2$, that were 5.1, 6.3, and 7.6, respectively, considering the values of the ideal Knudsen selectivities of 2.8, 3.7, and 4.7 for the corresponding binary gases. Freestanding MOF-TFs are also achievable through this type of synthesis strategy [162] (Figure 10C). The self-sacrificing templated synthesis strategy is also able to be employed with other types of synthesis strategies, which broadens its application in the fabrication of MOF-TFs. Schäfer et al. [163] fabricated MOF-TFs via the ECD through a self-sacrificing synthesis strategy. HKUST-1 TFs were obtained on copper foils only when surface oxide layers were present on the electrode. Furthermore, only Cu$_2$O was found to be working in the formation of HKUST-1, not CuO. The hypothesis is that the active form of copper precursor for HKUST-1 in the ECD system was Cu$_2$O from the oxidation of copper foil.

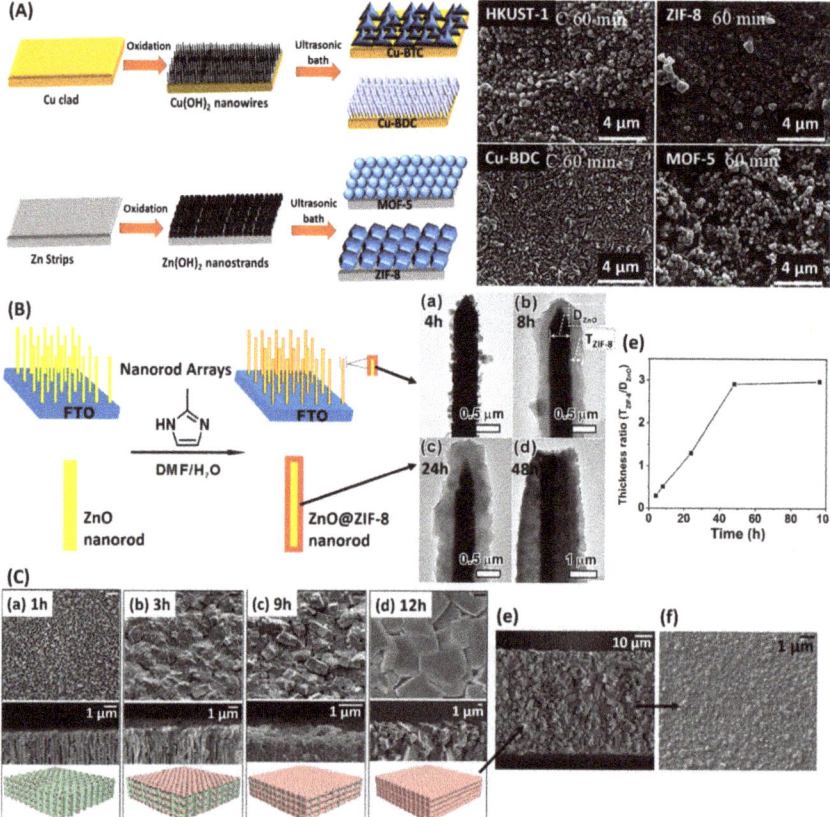

Figure 10. (**A**) Schematic of the two-step method for the fabrication of MOF-TFs, and SEM images of different types of MOF-TFs after sonication for 1 h. (Reproduced with permission from Abuzalat et al., Ultrasonics sonochemistry; published by Elsevier BV, 2018.) (**B**) Schematic of ZnO@ZIF-8 NRs synthesized

via the self-template strategy, and (**a–d**) TEM images of the NRs obtained after reaction for a different time and (**e**) the thickness ratio (T_{ZIF-8}/D_{ZnO}) in the NRs as a function of the reaction time. (Reproduced with permission from Zhan et al., Journal of the American Chemical Society; published by American Chemical Society, 2013.) (**C**) (**a–d**) Top-view and cross-section SEM images and simulation models of four states of the membrane representing the reaction process as a function of time; green-anodic aluminum oxide, pink-MIL-53 MOF. (**e,f**) The cross-section SEM images of the whole freestanding membrane. (Reproduced with permission from Zhang et al., Scientific reports; published by Nature Research, 2014.)

Metal oxides and hydroxides with high reactivity (i.e., they are easily ionized in solution) are preferable in this type of synthesis strategy, where the metal dissolution process is a crucial factor. Some commonly used metal oxides are Cu_2O, ZnO, and Al_2O_3, and hydroxides $Zn(OH)_2$, $Cu(OH)_2$, $Ni(OH)_2$, $Ca(OH)_2$, $Mg(OH)_2$, and $Al(OH)_3$. Considering the diverse and accessible macro-, micro-, and nanomorphology of metal, metal oxides, and metal hydroxides, many MOF-TFs, patterns, and composite structures can be made via self-sacrificing templated synthesis based on a suitable metal source.

5. Other Types of General Synthesis

Although the previously reviewed fabrication techniques based on traditional solution-processing techniques have been extensively studied and widely used in the fabrication of MOF-TFs, the limited scalability of these solution-based approaches prevent their large-scale production. Several common disadvantages of the solvent-processing fabrication techniques for MOF-TFs are (1) the large amount of solution needed due to the low volumetric yield of TFs, (2) the difficulty in preventing the formation of bulk MOFs and controlling the waste generation, and (3) potential safety and processing issues when using solvents, especially under solvothermal conditions. Therefore, many other types of synthesis strategies have been developed to realize a more economical and greener process. In this section, we will review currently reported fabrication techniques based on the solid–solid synthesis, vapor–solid synthesis, vapor–gel synthesis, and the special post-assembly method.

5.1. Solid–Solid Synthesis

As a solution to overcome the difficulties of solution-processing techniques, fabricating MOF-TFs in a solvent-free manner becomes an emerging strategy that can realize green synthesis [164]. Employing metal oxides or hydroxides instead of typically conducted metal salts in the reaction with organic linkers eliminates the production of acids since this type of synthesis strategy involves simple acid-base neutralization with water as the only byproduct.

Stassen et al. [164] demonstrated the fabrication and patterning of ZIF-8 TFs via solvent-free synthesis through the reaction of ZnO with melted 2-mIm (Figure 11A). This study obtained a dense ZIF-8 TF on Si wafer by sputtering, a patterned ZnO TF by μCP, and a flake-like ZnO TF on a carbon steel support by ECD. After covering with a thin layer of finely ground 2-mIm powders, the conversion of ZnO to ZIF-8 was completed at 160 °C in an oven, which also allowed for the evaporation of byproduct that was water. The crystal growth can be controlled by tuning the reaction time and changing the amount of prepared ZnO. The resulting ZIF-8 TFs precisely follow the original morphology of the ZnO TFs. Therefore, based on many morphologies of metal oxide TFs that can be obtained on different substrates using various techniques, the solvent-free synthesis strategy provides an easy route to apply many types of MOF-TFs on integrated devices in electronics and optics. However, this strategy is only preferable for MOFs that consist of organic linkers with a low melting point.

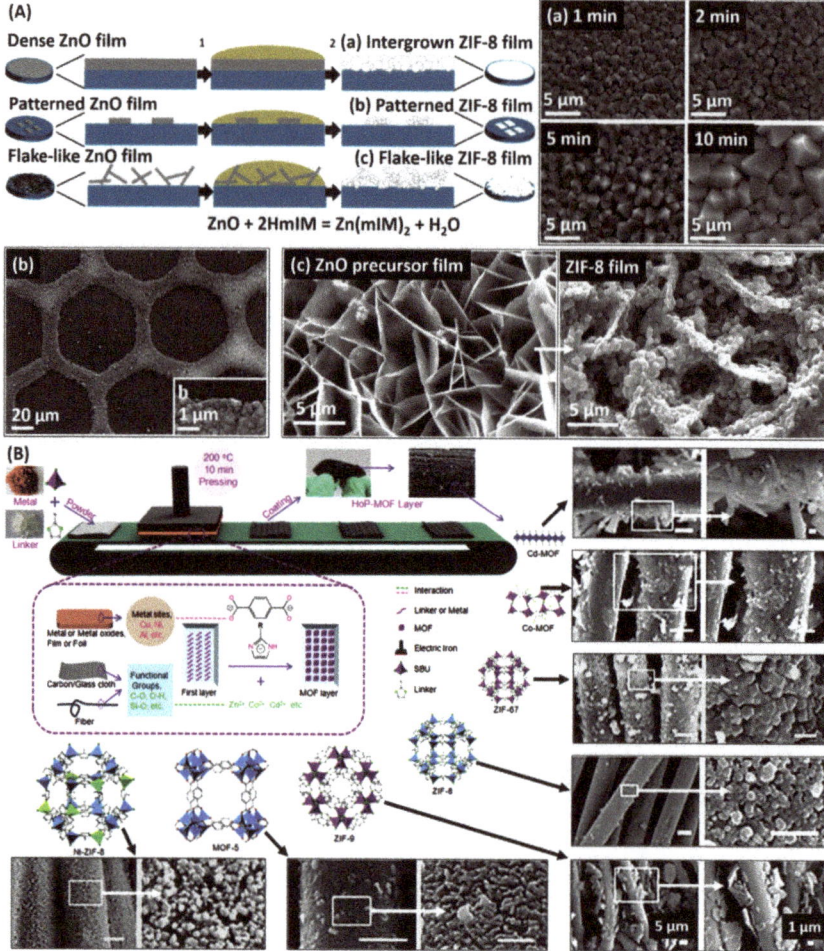

Figure 11. (**A**) Schematic of the solvent-free ZIF-8 film processing and patterning approach. SEM images of (**a**) intergrown ZIF-8 TFs at different times during the transformation of 1-μm thick sputtered ZnO film on Si wafer, (**b**) a ZIF-8 pattern obtained after 20-min transformation of a ZnO pattern, and (**c**) an electrochemically deposited flake-like ZnO precursor film and resulting ZIF-8 film after a 20-min transformation. (Reproduced with permission from Stassen et al., CrystEngComm; published by Royal Society of Chemistry, 2013.) (**B**) Schematic presentation of the hot-pressing method for MOF-TFs, and the overview (all scale bars, 5 μm) and enlarged (all scale bars, 1 μm) SEM images of different MOF-TFs fabricated on carbon cloth. (Reproduced with permission from Chen et al., Angewandte Chemie International Edition; published by Wiley-VCH, 2016.)

Following the concept of solvent-free synthesis, Chen et al. [165] introduced a hot-pressing (HoP) synthesis strategy to fabricate MOF-TFs under binder-free and solvent-free conditions (Figure 11B). In a general HoP process, both metal and organic precursors of MOFs are placed on a substrate and are pressed by a heating source, such as an electric iron, where the applied heat triggers the reaction between the precursors and leads to the formation of MOF-TF on the substrate. Different MOF-TFs, including MOF-5, ZIF-8, ZIF-9 (Co(bIm)$_2$), ZIF-67 (Co(2-mIm)$_2$), Co-MOF (Co(dcIm)$_2$, where dcIm = 4,5-dicyanoimidazole), Cd-MOF (Cd(Im)$_2$, where Im = imidazole), and bimetallic Ni-ZIF-8, have been realized on various flexible substrates, including carbon cloth, AAO film, nickel foam, copper foil, glass cloth, and glass fiber, via the HoP strategy. Take the fabrication of ZIF-8 TFs on carbon cloth as an example. Zinc acetate, 2-mIm, and polyethylene glycol (PEG) were manually ground and mixed. The mixture was then loaded on a 2 cm × 2 cm carbon cloth, packed with aluminum foil, and heated with an electric heating plate at 200 °C for 10 min. After peeling off the aluminum foil, the TF was washed with ethanol and DMF (each for 1 h) and stored in ethanol. All the ZIF-8 TFs obtained on different substrates presented uniformly distributed crystals. The HoP strategy can realize the fabrication of MOF-TFs on flexible materials with a controllable scale, which could be of interest in flexible electronics. However, this method requires high thermal stability of the MOFs and substrates.

5.2. Vapor–Solid Synthesis

A vapor–solid synthesis method is promising to apply MOF-TFs on devices that are incompatible with wet-based processing due to the risk of corrosion and contamination. There are several ways to achieve the fabrication of MOF-TFs in a vapor–solid synthesis manner, such as atomic layer deposition (ALD), chemical vapor deposition (CVD), and physical vapor deposition (PVD).

As mentioned previously, the ALD technique is widely used in seed-assisted secondary growth for MOF-TFs and in the preparation of metal precursor TFs and patterns. Other than these, ALD can also be directly employed to fabricate MOF-TFs via the vapor–solid reaction. In a general TF fabrication process via ALD, metal and organic precursors for the desired MOFs are sequentially blown onto the surface of a substrate with a pulse of gas to form the TF. Lausund and Nilsen [166] reported the deposition of UiO-66 ($Zr_6O_4(OH)_4(BDC)_6$) TFs via the ALD technique (Figure 12A). In this study, an amorphous organic–inorganic hybrid TF based on $ZrCl_4$ and H_2BDC formed on Si wafers. The vaporization temperatures for $ZrCl_4$ and H_2BDC were 165 and 220 °C, respectively. Then, the TF crystallized into a UiO-66 structure by heating to 160 °C for 24 h in a sealed autoclave with about 0.1 mL of acetic acid vapor. Uniform and continuous UiO-66 TFs were obtained locally on the wafer, acquiring a relatively low surface roughness. This ALD-based synthesis strategy provides an easy way to control the film thickness at an atomic level and is also available in the fabrication of multilayer structures. However, there are a few obstacles. First, the precursors, especially the organic precursors, must be volatile but not be subject to unwanted decomposition during the vaporization. Then, the capital cost of ALD equipment is high, and the process of ALD is slow, which prevents broad application of this technique.

The CVD technique is another type of vapor-based synthesis strategy in the fabrication of MOF-TFs with precisely controlled thicknesses. There are two steps in a general CVD process for fabricating MOF-TFs, including a metal oxide deposition step and a vapor–solid reaction step (Figure 12B). Stassen et al. [167] developed the concept of a CVD-based synthesis strategy for MOF-TFs. In this study, the ZnO layer was deposited first via ALD, and then the 2-mIm organic linker vapor was introduced to the reaction system via CVD, undergoing a metal oxide-to-MOF conversion process similar to the previously mentioned liquid/solid–solid synthesis. When fully converting 3 and 6 nm ZnO precursor layers, it resulted in ZIF-8 TFs with an average thickness of 52 and 104 nm, respectively. All the resulting ZIF-8 TFs achieved a uniform and controllable thickness. This CVD approach also realized lift-off patterning and fabrication of MOF-TFs on delicate features.

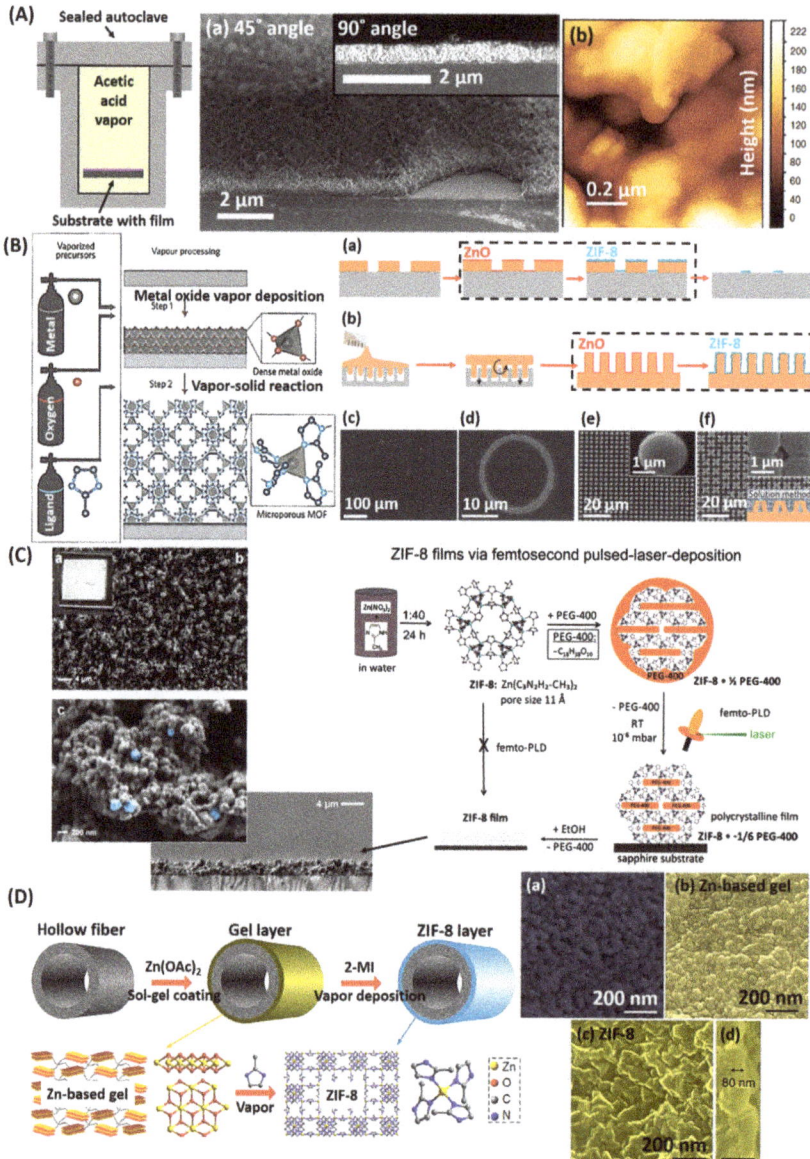

Figure 12. (**A**) Experimental setup for the heat treatment in acetic acid vapor. (**a**) Cross-section SEM images of the UiO-66 TF after treatment in acetic acid vapor, and (**b**) AFM image of the same film. (Reproduced with permission from Lausund et al., Nature communications; published by Nature Research, 2016.) (**B**) Schematic of the CVD process for ZIF-8 TFs (zinc-grey, oxygen-red, nitrogen-light blue, and carbon-dark blue). Schematics of (**a**) ZIF-8 pattern-deposition by MOF-CVD and subsequent lift-off of a patterned photoresist and (**b**) the production of ZIF-8-coated PDMS pillars by soft lithography

and MOF-CVD. SEM images of (**c,d**) the manufactured ZIF-8 patterns, (**e**) MOF-CVD-coated PDMS pillars, and (**f**) identical PDMS pillars after conventional solution processing of ZIF-8. The MOF-CVD processing steps are indicated with a dashed line in a and b; oxide and MOF films are represented in red and blue, respectively. (Reproduced with permission from Stassen et al., Nature materials; published by Nature Publishing Group, 2016.) (**C**) Schematic of ZIF-8 film preparation via the femto-PLD technique, and SEM images of PEG@ZIF-8 films on sapphire films with an optical image inserted. Crystals showing the typical ZIF-8 morphology are highlighted in light blue. (Reproduced with permission from Fischer et al., Chemistry of Materials; published by American Chemical Society, 2017.) (**D**) Schematic of the GVD fabrication process of ultrathin ZIF-8 film. SEM image top view of (**a**) a PVDF hollow fiber, (**b**) a Zn-based gel layer, and (**c**) a ZIF-8 TF with (**d**) a cross-section image. (Reproduced with permission from Li et al., Nature communications; published by Nature Research, 2017.)

Considering the extensive studies in the fabrication of MOF-TFs via various chemical means, there is a rare investigation related to physical deposition. Concerning this, Fischer et al. [168] developed a femtosecond pulsed-laser deposition (femto-PLD) technique to achieve the fabrication of MOF-TFs, extending the fabrication of MOF-TFs to PVD (Figure 12C). In the study, ZIF-8 powder was stabilized in PEG-400 and then pressed into a pellet to prepare the target for femto-PLD. PEG@ZIF-8 TF was deposited on a sapphire substrate at room temperature by ablating the target in a high-vacuum system for 6 h. The femtosecond laser with a wavelength of 516 nm at 442 fs was operated with a laser power of 30 mW (energy per pulse of 0.03 mJ at 1 kHz), and the laser beam was focused on the target surface with a spot size of 0.05 mm. This method produced mesoporous ZIF-8 films consisting of nanosized ZIF-8 crystals as shown in the SEM images in Figure 12C. The limitation of this femto-PLD strategy is that it is applicable only to the fabrication of highly porous and thermally labile MOF-TFs with stabilizing additives.

5.3. Gel–Vapor Synthesis

Li et al. [169] introduced a gel–vapor deposition (GVD) strategy for the fabrication of MOF-TFs (Figure 12D). This method combines sol-gel coating with vapor deposition to receive the benefits from both techniques and realizes solvent-/modification-free and precursor-/time-saving synthesis of MOF-TFs with a controlled thickness. In this study, a Zn-based sol was prepared by mixing $Zn(CH_3CO_2)_2 \cdot 2H_2O$ and ethanolamine in ethanol and then coated on ammoniated polymer hollow fibers. The formation of Zn-based gel was initiated by heat treatment; the gel was then subjected to the conversion process of gel-to-MOF by introducing a vapor of organic linkers. An ultrathin ZIF TF thinner than 20 nm was obtained by adjusting the sol concentration and coating procedure. Unlike the liquid–solid reaction, the GVD process based on gas–solid reactions does not involve the diffusion of precursors in solutions and thereby offers better control over the reactant transport and fluid dynamics during crystallization. The GVD process requires no pretreatment of substrates and can be applied to many types of substrates. However, the MOF-TFs prepared via this strategy may face a problem of shrinkage and cracking during drying.

5.4. Post-Assembly Method

Most synthesis strategies to fabricate MOF-TFs use separate metal and organic linker precursors; the process often involves a chemical reaction step and a framework assembly step. There is a different strategy, using preformed MOF particles on substrates, to achieve MOF-TFs, which involves only the framework assembly step. Building on the extensive knowledge learned about the synthesis of MOF particles, MOF-TFs can be fabricated via a post-assembly process through a variety of strategies.

The Langmuir–Blodgett (LB) deposition method is a well-established technique for the fabrication of orderly monolayers on liquid–substrate surfaces, which is often applied using the LBL deposition strategy to assemble preformed MOF NCs for the formation of dense MOF-TFs. Makiura et al. [170] demonstrated the fabrication of MOF-TFs via the LB-LBL deposition strategy (Figure 13A). In this study, organic CoTCPP (tetrakis(4-carboxyphenyl)-porphyrin-cobalt^{II}) and pyridine were spread onto an aqueous solution containing $CuCl_2$ and formed a layer of CoTCPP-pyridine-Cu (NAFS-1) MOF nanosheets on the solution surface. The MOF monolayer was compressed by the container's walls and received proper surface pressure. Then, the layered MOF nanosheets were transferred onto a smooth surface (Si wafer or quartz substrate) via the LB method. The process was repeated several times in an LBL manner to obtain an interdigitated NAFS-1 MOF-TFs. MOF-TFs prepared by the LB-LBL deposition strategy show the low surface roughness and high homogeneity, and the film thickness can be adjusted by the number of deposited layers and is accurate to the nanometer level [170–177]. Moreover, the choice of substrates for this type of MOF-TF deposition strategy is widely open. However, due to the weak interactions (p stacking or Van der Waals' force) between the film and the substrate, MOF-TFs prepared by the LB method may easily fall off. Furthermore, the requirement of a generally expensive LB device for the deposition largely limits its application.

Another similar strategy, the Langmuir–Schäfer (LS) deposition, has also been applied to fabricate thin layers of MOFs based on preformed MOF NCs. Wang et al. [178] reported the fabrication of a series of M-TCPP(Fe) (M = Co, Cu, and Zn) MOF-TFs via the LS deposition strategy (Figure 13B). Unlike the LB deposition method, where the substrate is slowly immersed into and then withdrawn from the surface, yielding a mono-/multilayered TF, the substrate in the LS method is situated in a horizontal position and slowly put in contact with the liquid surface and subsequently lifted off the liquid surface. In this study, 2D M-TCPP(Fe) nanosheets were first prepared via a surfactant-assisted solvothermal synthesis, followed by the deposition onto substrates by the LS deposition process to obtain mono-/multilayers. Take the deposition of Co-TCPP(Fe) as an example. First, $Co(NO_3)_2$, pyrazine, and polyvinylpyrrolidone were dissolved in a mixture of DMF and ethanol (V:V = 3:1), and then mixed with TCPP(Fe) in another mixture of DMF and ethanol (V:V = 3:1). After the solution was sonicated for 10 min, the mixture was set to 80 °C for 24 h. The resulting dark brown products were washed twice with ethanol and collected by centrifuge. Finally, the obtained Co-TCPP(Fe) nanosheets were dispersed in ethanol to obtain a colloidal suspension with a concentration of 1.0 mg/mL. Then, the suspension was gently dropped onto the surface of water in a beaker. After Co-TCPP(Fe) nanosheets spontaneously spread to form a TF on water, the film was transferred onto a solid substrate via the LS method. Finally, the TF-coated solid substrate was immersed into fresh water to remove the loosely deposited nanosheets and then blow-dryed with N_2. By this step, one deposition cycle was completed. Similar to the LB method, the LS method could follow an LBL deposition manner to control the thickness of the resulting MOF-TFs. The resulting Co-TCPP(Fe) MOF-TF with five deposition layers showed the best electrochemical catalytic activity toward the reduction of H_2O_2, with a limit of detection of 0.15×10^{-6} mol/L. The developed sensor was available for real-time detection of H_2O_2 secreted by live cells.

Xu et al. [179] reported a modular assembly strategy based on LS deposition to prepare MOF-TFs (Figure 13C), which realized a rapid fabrication of oriented MOF-TFs with larger domain sizes than the ones in the LB method [170]. Cu-TCPP nanosheets were first synthesized via a solvothermal reaction of $Cu(NO_3)_2$ and H_2TCPP. The formed MOF nanosheets were first dispersed in ethanol to form a colloidal suspension and then added onto the surface of the water in a beaker dropwise to create a TF of Cu-TCPP MOF. The MOF-TF was then transferred to a quartz substrate via stamping. Repeating the stamping process in an LBL manner increases the film thickness. A highly oriented MOF-TF of 100 layers can be prepared within 10 min, enabling the preparation of oriented MOF-TFs in a shorter time than many previously reviewed fabrication techniques, although the orientation of the resulting TF depends only on the orientation of nanosheets obtained from bulk MOF synthesis. An obstacle in widening the application of the LS strategy is, similar to the LB strategy, the requirements of preformed MOF crystals with high aspect ratios can obtain a uniform air–liquid interface for MOF-TF deposition, which remains a significant challenge in the synthesis of MOFs.

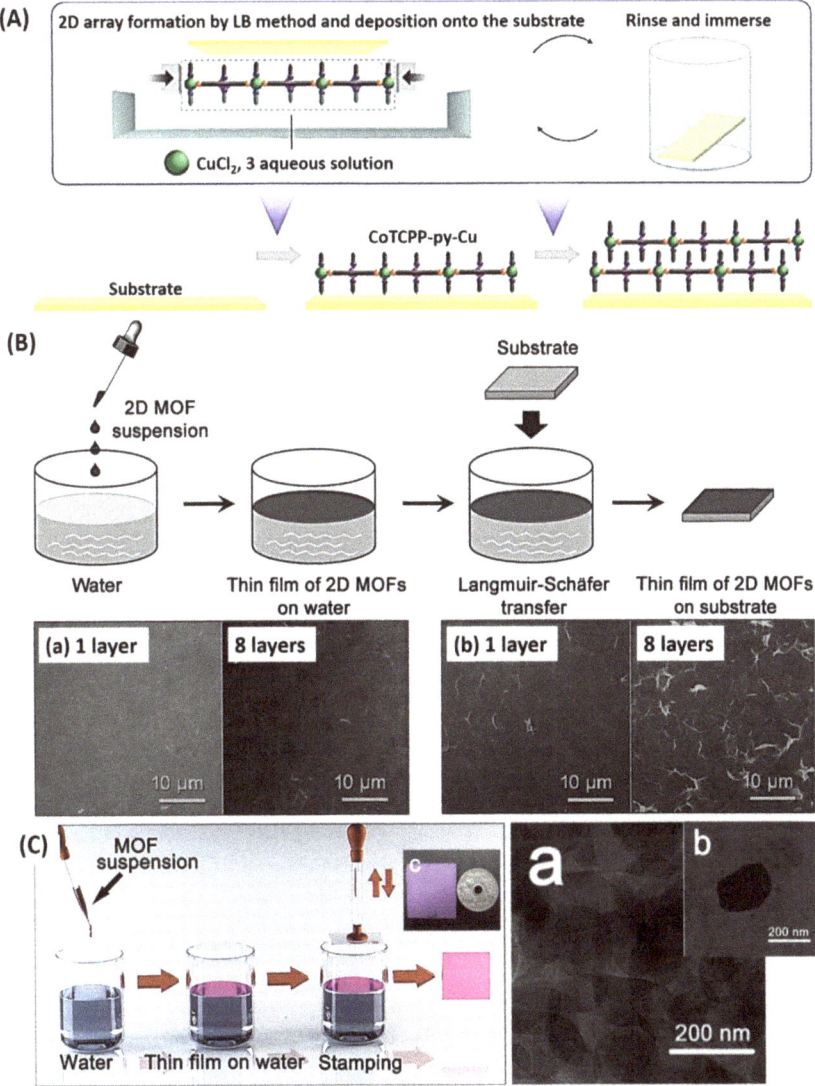

Figure 13. (**A**) Schematic illustration of the fabrication method of NAFS-1 MOF-TF via the LB-LBL deposition strategy (C-grey, N-blue, O-red, Co^{2+}-pink, and Cu^{2+}-green). (Reproduced with permission from Makiura et al., Nature materials; published by Nature Publishing Group, 2010.) (**B**) Schematic illustration of the assembly process for the preparation of 2D-MOF-nanosheet-based TFs. SEM images of (**a**) Co-TCPP(Fe) and (**b**) Cu-TCPP(Fe) MOF-TFs on Si wafers via the LS method with different deposition cycles. (Reproduced with permission from Wang et al., Advanced Materials; published by Wiley-VCH, 2016.) (**C**) Illustration of the assembly process of this MOF-TF. (**a,b**) TEM images of the synthesized Cu−TCPP nanosheets, and (**c**) an optical photo of the MOF-TFs after 15 deposition cycles on a quartz substrate. (Reproduced with permission from Xu et al., Journal of the American Chemical Society; published by American Chemical Society, 2012.)

Other deposition strategies, such as SC [121] and the drop-casting method [121], have also been applied to assemble preformed MOF crystals on substrates; however, the resulting film morphologies were not as good as the ones made by LB or LS deposition.

6. Modification of MOF-TFs

For practical application aspects of view, such as gas sensing, a real gas environment is often a multi-component system. Consequently, selective adsorption of a specific molecule from a gas mixture is desired. Although MOF-TFs, by nature, provide excellent selective adsorption over many chemical species at the molecular level, their selectivity still cannot meet the requirements in practical applications [180]. Besides, many MOF-TFs have inadequate proton-conductivity capacities and are preferable for fabrication on conductive substrates for electronic gas sensing [181], but the inorganic substrates are relatively expensive. Although promising in lowering the cost, the fabrication of MOF-TFs on porous polymers is still limited by techniques because of the poor combination force between substrates and MOF-TFs [135]. Therefore, part of the research interest is shifting to the modification of MOF-TFs, such as modifying the macro-/micro-structures and tuning the pore functionality. The modification of MOF-TFs can be conducted on, but not limited to, ligands, metal sites, pores, and the overall hierarchical structure.

The modification of MOF-TFs can be classified into two categories, i.e., the in situ modification during synthesis, and the post-synthesis modification.

6.1. In Situ Modification

Mao et al. [182] developed a modification strategy to functionalize MOF-TFs during the fabrication process based on the functionalization of self-sacrificing metal hydroxide templates. In the study, copper hydroxide nanostrands (CHNs) were first mixed with different negatively charged functional species, including Au NPs, $[AuCl_4]^-$ anions, ferritin, and glucose oxidase (GOx), polystyrene spheres, and single-walled carbon NTs (SWCNTs), and then filtered onto porous substrates. Followed by a 1-h reaction with H_3BTC solution at room temperature, HKUST-1 composite TFs with encapsulated functional components were obtained. The resulting HKUST-1 TFs that encapsulated with different functional species exhibited interesting hybrid functions. The GOx/HKUST-1 composite TF showed interesting enzymatic activity toward glucose. The Au NP/HKUST-1 composite TF presented an excellent catalytic performance in the oxidation of CO, with a catalysis efficiency of 297.62×10^{-6} mol/g_{Au}s at 160 °C based on 50% conversion. The SWCNT/HKUST-1 composite TF possessed impressive conductivity and flexibility, showing a large BET surface area of 1192 m^2/g and high thermal decomposition temperature of 333 °C. This synthesis strategy offers a route to encapsulate functional species into an MOF system, obtaining synergistic and size-selective functional materials. LBL deposition is capable of encapsulating guest molecules into the structure of MOF-TFs. Gu et al. [183] reported in situ LBL growth of lanthanide coordination compound-encapsulated MOF-TFs using a modified LPE pump method (Figure 14A). In this study, the $Ln(PDC)_3$ (PDC = pyridine-2,6-dicarboxylate; Ln = Eu, Tb, and Gd) was encapsulated in the pores of MOFs during the fabrication process for HKUST-1 TF via the LBL deposition strategy. The resulting composite TFs showed RGB (red, green, and blue) primary colors because of the loading of $Ln(PDC)_3$. Moreover, the proportion of different $Ln(PDC)_3$ in the mixture can be adjusted to obtain a TF of white light emission. This synthesis strategy was useful in the development of oriented and homogeneous solid-state lighting composite TFs with a high encapsulation efficiency. Later, Fu et al. [184] employed this synthesis strategy and obtained a trans-azobenzene@HKUST-1 hybrid TF by immersing the substrate in $Cu(CH_3COO)_2$ solution, H_3BTC solution, and trans-azobenzene solution, sequentially. The resulting hybrid MOF-TF possessed photo-switching and photoluminescent properties based on the trans-azobenzene, and a temperature-dependent photoluminescent emission, leading to a new path for the preparation of photochromic TFs and the development of multifunctional optical devices and sensors.

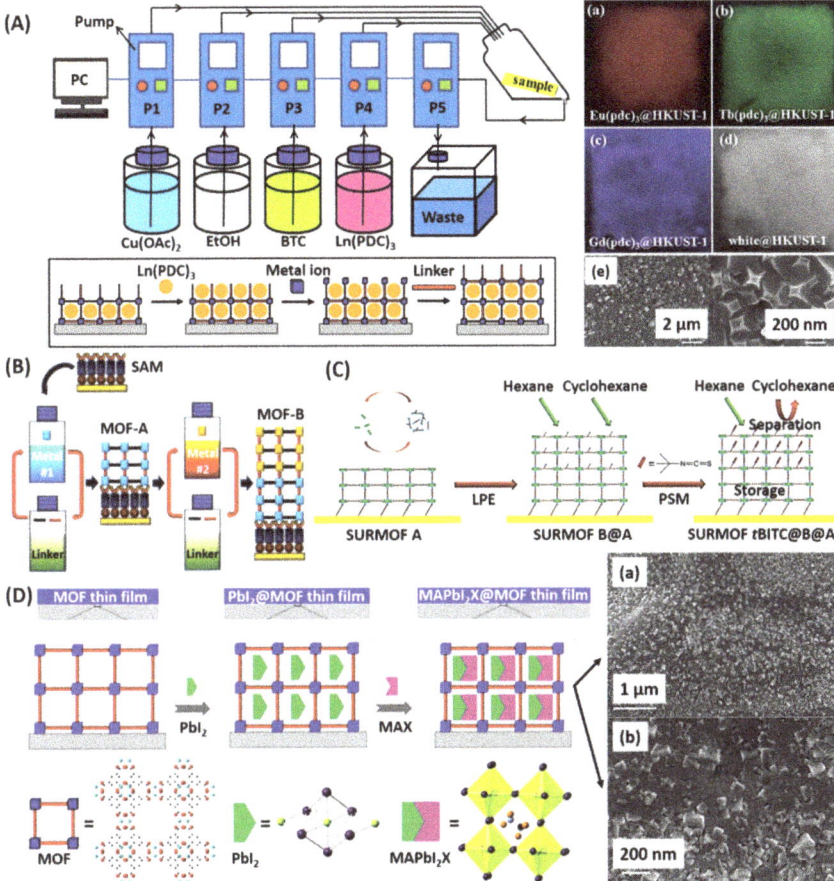

Figure 14. (**A**) Schematic presentation of in situ LBL growth of Ln(PDC)$_3$-encapsulated HKUST-1 TF using a modified LPE pump method. Photographs of a UV-irradiated (**a**-**c**) Ln(PDC)$_3$@HKUST-1 TF and (**d**) a mixed Ln(PDC)$_3$@HKUST-1 film on quartz glasses via the modified LPE pump method, and (**e**) SEM images of the Eu(PDC)3@HKUST-1 TF. (Reproduced with permission from Gu et al., ACS applied materials & interfaces; published by American Chemical Society, 2015.) (**B**) The LBL method for the hetero-epitaxial growth of MOF-on-MOF hybrid TF structure on SAMs. (Reproduced with permission from Shekhah et al., Dalton Transactions; published by Royal Society of Chemistry, 2011.) (**C**) Schematic illustration of programmed functionalization of SURMOFs via hetero-LPE growth and the PSM process. (Reproduced with permission from Tu et al., Dalton Transactions; published by Royal Society of Chemistry, 2013.) (**D**) Schematic illustration of a confined synthesis of MAPbI$_2$X (X = Cl, Br, or I) in the interior pores of oriented MOF-TF, and (**a**,**b**) SEM images of MAPbI$_2$Br@HKUST-1 TF. (Reproduced with permission from Chen et al., ACS applied materials & interfaces; published by American Chemical Society, 2016.)

Another way of in situ modification for MOF-TFs is the stratified synthesis strategy, which enables MOF-on-MOF heterostructures with hierarchical porosity. Shekhah et al. [115] demonstrated the feasibility of the LBL deposition strategy in the stratified synthesis for MOF-on-MOF TFs (Figure 14B). First, SAMs were deposited on Au/Ti/Si wafers or Au substrate in an ethanolic solution of 4,(4-pyridyl)phenyl-methanethiol. Ethanolic solutions of zinc acetate or copper acetate and an equimolar H$_2$NDC/DABCO (DABCO = 1,4-diazabicyclo(2.2.2)octane) mixture were used as precursors for each

MOF. The MOF-TFs were fabricated via the traditional LBL deposition method, in which the substrates were first immersed in metal acetate solution for 30 min and then in the H$_2$NDC/DABCO solution mixture for 60 min, with a washing step with pure ethanol of 5 min in between. The MOF-on-MOF fabrication was completed by 60 LBL cycles of first MOF-TF and then 60 LBL cycles of the second MOF-TF. In the end, highly oriented Zn-MOF-on-Cu-MOF hybridized TFs were obtained on SAM-based substrates, showing precisely controlled thicknesses, a result of the LBL deposition. Many different MOFs were studied using stratified synthesis, including MOF-on-MOF heterostructures [116] and MOF@MOF core-shell hybrid TFs [117].

6.2. Post-Synthesis Modification

Other than the in situ modification during fabrication, the functionality of MOF-TFs can be modified by a post-synthesis modification (PSM) process.

Shekhah et al. [185] used layer-pillar-based SURMOFs with BDC-NH$_2$ linker as a robust platform to attach ferrocene via a PSM process. First, the SURMOF was synthesized on an SAM-modified Au(111) surface terminated with -OH groups from 11-mercaptoundecanol. The layer-pillar-type SURMOF consisted of Cu$_2$(BDC-NH$_2$)$_2$ MOF nanosheets connected by DABCO pillars. 1-ferrocenylmethy-lisocyanate was used as the PSM reagent. To perform PSM, a pristine SURMOF was immersed in a solution of 1-ferrocenylmethy-lisocyanate dissolved in dichloroethane at 30 °C. After the PSM, the SURMOF could provide a maximum density of two ferrocene molecules per pore. The encapsulation of ferrocene inside the pores would alter the electrochemical properties of the SURMOFs. Based on the size-selective adsorption property of MOFs, many small molecules, such as redox-active compounds [186] and fluorescent dye molecules [187], can be encapsulated in MOF pores for integrated properties. Partial linker exchange can also be achieved via the PSM by tuning the reaction conditions [188].

Based on an MOF-on-MOF heterostructure via stratified synthesis, further modification of the pore functionality is available through a PSM process. Tu et al. [189] fabricated an SURMOF-on-SURMOF hybrid TF on the pyridyl-terminated Au-covered QCM substrate via LPE growth (Figure 14C). The fabrication of SURMOFs followed the LBL deposition manner, including 5-min immersion in copper acetate solution, 10 min in BDC/DABCO solution, and 5 min in ethanol for washing purposes between the two precursors. Before PSM, the SURMOFs were kept in dichloromethane overnight, followed by treatment under vacuum overnight at room temperature. Then, the activated samples were exposed to the vapor of *tert*-butyl isothiocyanate (*t*BITC) under static vacuum at room temperature for two days. Afterward, the modified SURMOFs were treated under a dynamic vacuum at room temperature for 1 day to remove physically adsorbed *t*BITC molecules. Through the PSM, the pore size of the top SURMOF, Cu$_2$(BDC-NH$_2$)$_2$(DABCO) MOF was modified by targeting the amino groups with *tert*-butyl isothiocyanate (*t*BITC). This hybrid TF with a modified surface layer presented extraordinary separation and storage of hexane over cyclohexane according to the unique size selectivity from the small pore size of the outer SURMOF achieved via the PSM process and the high storage capacity from the inner SURMOF without modification. Chen et al. [190] conducted an oriented MOF-TF via the LPE approach as a host for perovskite QD fabrication (Figure 14D). By introducing PbI$_2$ and CH$_3$NH$_3$X (MAX; X = Cl, Br, and I) precursors into an HKUST-1 TFs, perovskite MAPbI$_2$X QDs were synthesized in the pores of the HKUST-1 TF with uniform diameters of 1.5–2 nm, which matched the pore size of HKUST-1. This perovskite QD@MOF-TFs was shown to be stable to moist air with 70% humidity for 4 days. This strategy provides confined synthesis of perovskite QDs with high stability and a uniform size distribution according to the oriented MOF-TF, broadening the application of perovskite QDs in luminescent sensors, photovoltaic solar cells, and other optoelectronic devices.

7. Patterning of MOF-TFs

There are several ways to prepare patterned MOF-TFs following previously discussed synthesis strategies. The patterning strategies can be grouped into three categories, i.e., bottom-up patterning of the substrate surface, top-down patterning of the substrate surface, and patterning of templates.

Patterned substrate surface via the bottom-up strategy can be realized by patterning functional groups on the substrate surface that either supports the growth of MOFs or prohibits it on desired areas. µCP is the most popular technique to pattern SAM layers for the fabrication of patterned MOF-TFs [59,107,125,191]. Precise SAM patterns (e.g., dots, squares, and lines), facilely prepared by the µCP technique, can result in accurately controlled MOF patterns. Besides SAMs, other alternatives can assist the patterned fabrication of MOF-TFs. Liang et al. [192] reported the fabrication of patterned MOF-TFs on substrates based on printed protein patterns (Figure 15A). Protein patterns of bovine serum albumin (BSA) were created on Si wafers and PET polymer films via stamping that transferred the aqueous BSA solution. Then, MOF precursor solutions were added onto the BSA patterns to form MOF crystals confined to the patterned areas. The patterning of MOFs can be achieved only on BSAs because BSAs can rapidly accumulate metal cations and organic ligands in solution, which accelerated the growth rate of MOFs on the pattered area. For the synthesis of ZIF-8 TFs and patterns, a fresh mixture of zinc acetate and 2-mIm was prepared in deionized water and transferred onto the BSA patterns. ZIF-8 patterns were obtained after 12 h. A similar process was performed for the fabrication of patterned $Ln_2(BDC)_3$ MOF-TFs and patterns. A fresh mixture of $LnCl_3$ (Ln = Tb, Eu, and Ce) and Na_2BDC was prepared in deionized water and transferred onto the BSA patterns. After 30 s of reaction, LMOF patterns were obtained. By this means, ZIF-8 and luminescent $Ln_2(BDC)_3$ MOF-TFs (including patterns) can be achieved on both rigid and flexible substrates.

Other fabrication techniques based on the bottom-up strategy are also available for MOF-TFs. As mentioned previously, ECD methods are quite accessible in the fabrication of patterned MOF-TFs. Li et al. [145] created desired patterns by writing PDMS on a conductive FTO substrate surface to fabricate patterned MOF-TFs via ECD. According to the insulating nature of PDMS, $Ln(OH)_3$ was only electrochemically deposited on the surface area without PDMS, thereby resulting in the formation of patterned Ln-MOF TFs on the exposed FTO glass surface. These LMOF-TFs and patterns are of interest in the fields of MOF-based biosensors, bio-medical devices, color displays, anti-counterfeiting, and potentially aids in crime scene investigation.

Patterned substrate surfaces can also be made via the top-down strategy. Navarro et al. [193] reported the preparation of patterned MOF-TFs based on a patterned substrate via laser ablation (Figure 15B). In this study, brass sheets were used as substrates and patterned physically by using laser irradiation to create lines, dots, and holes on the substrate. A 20 W Nd:YAG laser technique was used for ablation, which created different roughness for different areas on the substrate. Porous substrates with 1.6% and 18% porosities and 20-32 mm effective micro-perforation diameters were obtained. After creating different surface roughness, the formation of ZIF-8 TFs followed traditional solvothermal synthesis in an autoclave at 100 °C for 4–8 h, depending on different samples. Varied surface roughness would lead to different nucleation and growth rates of ZIF-8 TFs, which resulted in the patterning of ZIF-8 TFs. Other than this, the formed ZnO from Zn oxidation during laser irradiation was likely another key factor to promoting the preferred formation of ZIF-8 TFs on the laser-irradiated areas. The obtained ZIF-8 pattern on porous brass sheets exhibited excellent separation performance for H_2/CH_4 and He/CH_4 mixtures, showing separation factors of 14.4 and 8.7, respectively.

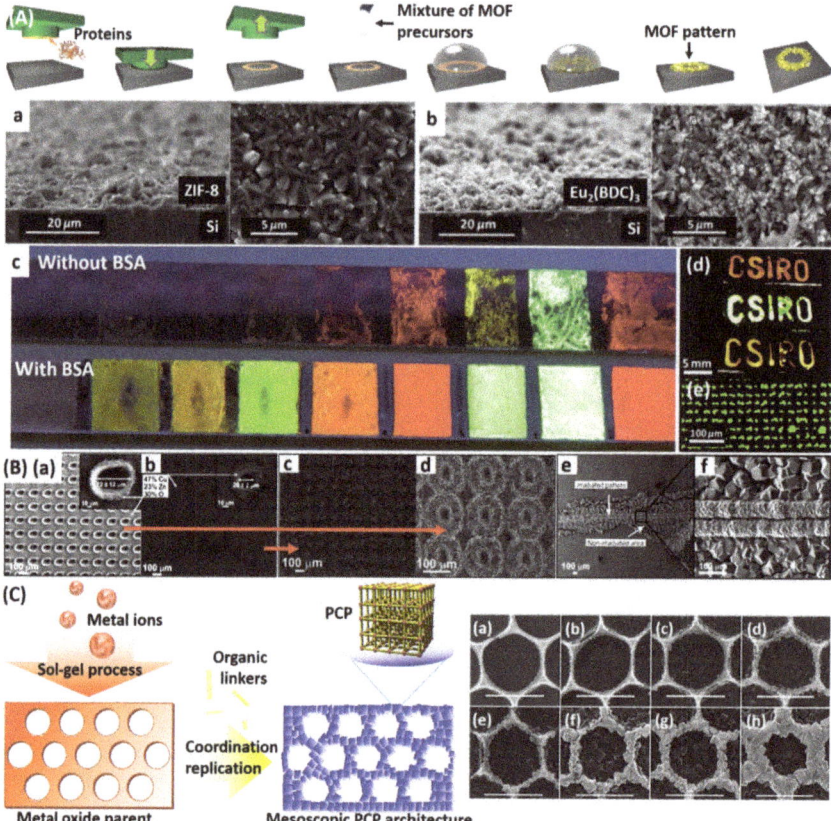

Figure 15. (**A**) Schematic showing the biomimetic replication of MOF patterns using a protein pattern. SEM images of (**a**) ZIF-8 and (**b**) Ln$_2$(BDC)$_3$ MOF-TFs via the biomimetic replication. Photograph under UV light of (**c**) Ln$_2$(BDC)$_3$ MOF-TFs formed with various mixing ratio of Eu, Tb, and Ce ions in the precursor solution; (**d**) Eu$_2$(BDC)$_3$ (red), Tb$_2$(BDC)$_3$ (green), and mixed Ln$_2$(BDC)$_3$ (yellow) patterns; and (**e**) Tb$_2$(BDC)$_3$ dot microarrays. (Reproduced with permission from Liang et al., Advanced Materials; published by Wiley-VCH, 2015.) (**B**) SEM images of (**a**) inlet (rough) and (**b**) outlet (smooth) sides of a laser-irradiated brass support (inserted enlarged perforations and EDX atomic compositions), and ZIF-8 TFs grown on (**c**,**d**) each side indicated by arrows, and on (**e**) linearly irradiated and (**f**) non-irradiated regions of the brass support. All scale bars, 100 µm. (Reproduced with permission from Navarro et al., Journal of Materials Chemistry A; published by Royal Society of Chemistry, 2014.) (**C**) The coordination replication and mesoscopic architecture concept, and SEM images of (**a**) the Al$_2$O$_3$ pattern and (**b**–**h**) the same sample after replication at 140 °C for 1, 4, 6, 10, 20, 40, and 60 s, respectively. All scale bars, 1 µm. (Reproduced with permission from Reboul et al., Nature materials; published by Nature Publishing Group, 2012.)

In addition, patterned MOF-TFs can be made by patterning self-sacrificing metal oxide or hydroxide templates. Reboul et al. [194] demonstrated the fabrication of patterned Al(OH)(NDC) MOF-TFs via self-sacrificing synthesis based on a patterned Al$_2$O$_3$ template (Figure 15C). The 2D/3D Al$_2$O$_3$ structures were prepared through a sol-gel process with polystyrene beads as the hard templates at 100 °C, which were removed through calcination at 580 °C for 7 h before the conversion of Al$_2$O$_3$ to Al(OH)(NDC) MOFs. The obtained Al$_2$O$_3$ templates were placed in a microwave container containing an aqueous H$_2$NDC solution, followed by a microwave treatment at 180 °C for 10 min. The microwave treatment was performed at 180 °C for 1 min. The microwave treatment to make mesoporous and macroporous structures of Al(OH)(NDC) MOFs was performed at 180 °C for 1 min. The resulting Al(OH)(NDC) MOF architectures with hierarchical porosity presented enhanced selectivity and mass transfer for water/ethanol separation due to the hydrophobic micropores of Al(OH)(NDC) and the mesopores/macropores inherited from the parent aerogels. The room temperature separation selectivity for water/ethanol was 2.26 for microporous Al(OH)(NDC) MOF crystals, 2.56 for the mesoporous structure, and 1.84 for the macroporous structure. It was suggested that the smaller pore size of the mesoporous structure and the higher exposed surface area of Al(OH)(NDC) MOF within this architecture allowed higher ethanol retention and improved separation selectivity. Other than using hard templates, photolithography is another way to direct the preparation of metal oxide templates to fabricate patterned MOF-TFs [195]. Moreover, hard templates can be applied in the interfacial synthesis of MOFs to achieve patterned MOF-TFs [128].

8. Conclusions

In conclusion, this review summarizes the current progress on the manufacturing of MOF-TFs, from the general fabrication techniques to the modification and patterning techniques. All the reported techniques in the synthesis of MOF-TFs were discussed, including their advantages and disadvantages, which can guide advancement in the fabrication processes of MOF-TFs to realize more controllable, scalable, and greener processing.

MOF-TFs, fabricated using different strategies, perform differently due to the differences in the resulting grain size, surface roughness, film thickness, film stabilities, etc. Although extensive studies have been carried out, the fundamental understanding of the MOF-TF formation process, including precursor reaction kinetics, surface reaction kinetics, reactant transport, film nucleation, and growth, is still at an early stage; more research is needed before the application of MOF-TFs at the industry scale. Considerable progress needs to be made to realize the controlled and green synthesis of MOF-TFs, especially for scale-up production, despite the emergence of many fabrication strategies for functional MOF-TFs.

Currently, direct synthesis and secondary growth strategies based on traditional hydro/solvothermal batch synthesis are still two of the most extensively used methods for their easy conduction. However, they possess a time-dependent growth rate for MOF-TFs, lending difficulty in controlling the fabrication process, which may result in an uncontrollable film thickness, and surface morphology. Moreover, the batch reaction often requires a high cost from large reactant consumption and waste production. Furthermore, there is a potential safety issue when using a high-pressure reactor. LBL deposition, ECD, DC deposition, and SC deposition are prevalent in the manufacturing of MOF-TFs and patterns. The LBL deposition strategy can realize the fabrication of many types of MOF-TFs and enable precise control of the amount and location/distribution of functionalities for tailored properties. It offers excellent control over the film thickness, gives rise to highly oriented and uniform TFs, and enables the fabrication of more complex MOF heterostructures. However, it shows its limitation on the fabrication of some types of MOF-TFs on specific substrates, and a slow growth rate for MOF-TFs.

Moreover, to achieve uniform films, the volume of precursor solutions used for the substrates to be immersed in is much larger than that of the substrate, contributing to a significant amount of solution wastage. The ECD strategy allows the rapid fabrication of defect (cracks and pinholes)-free MOF-TFs with controllable thicknesses. Relatively uniform and compact deposits can be made in template-based structures; ECD also has the advantages of higher deposition rates and inexpensive equipment due to the non-requirements of either a high vacuum or a high reaction temperature. However, it is limited to electrochemical deposition of non-conductive MOF-TFs on conductive substrates. Moreover, metal ions with high inertness could separate on the cathode, and the organic linkers may be oxidized.

Both the DC deposition and the SC deposition are simple methods to fabricate MOF-TFs. They can achieve thin uniform TFs with a wide range of thicknesses (from nm to μm). Both methods utilize small equipment that is less expensive than the setup in many other methods (e.g., ECD, HoP, and ALD), and they do not require high energy processes. These two techniques are ideal for coating MOF-TFs on flat substrates, not curved or flexible ones. Besides different operation procedures, there are several distinct features of these two techniques. First, the DC technique is available in large-scale TF manufacturing while the SC technique can only be applied to small substrates. Then, the drying time needed in DC deposition is significantly long while that in SC deposition is short, which may alter the film formation process due to the different times. Both methods result in high solution wastage. Similar to the LBL deposition, the DC deposition also requires a solution reservoir to immerse the substrate. The volume of the solution is much higher than the size of the substrate, leading to a high quantity of solution wastage. During SC deposition, the majority of the solution is cast off, resulting in high levels of solution wastage.

There are other types of fabrication techniques that have been reviewed that show an ability for controllable fabrication of MOF-TFs and patterns. Despite their novelty, there are many obstacles to broadening their application. For example, the HoP technique can realize the fabrication of various MOF-TFs on flexible materials with controllable scale, which could be of interest in flexible electronics. However, it requires high thermal stability of MOFs and substrates.

Considering the requirements of scale-up production with effective cost, the commercialization of MOF-TFs appears quite challenging. Despite the development of novel techniques to achieve green synthesis in terms of reduced chemical consumption and low waste production, which also significantly influences the production cost, it is hard to fulfill other requirements. For example, the ALD technique offers an easy way to control the film thickness at an atomic level and is available in the fabrication of multilayer structures. However, it only works on certain types of MOFs that use volatile but thermally stable organic precursors. Moreover, ALD equipment is expensive, and the process is slow, which increases the operation cost. ECD and DC methods are promising in realizing large-scale controllable manufacturing of MOF-TFs and patterns. However, they could not satisfy the strict requirements for low cost and green synthesis. Although the equipment used in these techniques could be relatively low, the ECD technique requires high energy input while the DC technique contributes to high levels of waste production. Currently, there is no ideal technique that can satisfy all the requirements for commercialization.

The recent advancements in the continuous-flow microreactor and process automation lead to new opportunities in advancing the progress on commercializing MOF-TFs. The synthesis of MOFs using systems involving microfluidics has been widely investigated [64]. Due to its small size, a microreactor can offer several advantages, including fast mixing of the reactants, an efficient way for mass and heat transfer, and precise control of the reaction dynamics. [196] The use of less liquid volume allows for greener synthesis and lowers the production cost. Besides, current progress in using automatic methods for MOF synthesis demonstrates promising results in the manufacturing of freestanding fine powders (including nanocomposites) [64,197] and membranes [64] on porous supports. However, there is no implementation of microfluidic-assisted techniques for the fabrication of MOF-TFs and patterns on the substrate surface. Coupling microreactors with current state-of-the-art deposition techniques to fabricate MOF-TFs could realize more exceptional control over synthesis conditions. It has the potential to enable more scalable, controllable, and greener synthesis in terms of controlling the TF formation process, shortening the fabrication time, reducing reactant consumption and waste production, and allowing for mechanical automatization amenable for scale-up. A wide range of film thicknesses (from nm to μm) and areas (from nm to cm) could be obtained on heterogeneous hierarchical TFs and patterns via microfluidic-assisted techniques. Moreover, it is known that the formation of highly ordered MOF crystals occurs first via an assembly of primary building blocks to define the secondary building blocks (SBUs) and then to the MOF crystallites [198]. Therefore, SBUs are the ideal seeds for the growth of MOF crystals [199,200]. However, the SBU-assisted secondary growth strategy is currently limited to the fabrication of bulk MOFs. Yet, it is promising in supporting TF fabrication without introducing foreign elements in the desired MOF structures.

Although extensive investigations and significant progress have been made for MOFs, due to their enormous chemistry possibility, there remains considerable room for future exploration and advancement. This review summarized and analyzed existing manufacturing techniques for MOF-TFs, including patterning, offering direction to researchers for the next generation development of MOF-TF processing, and building a foundation to realize the commercialization of MOFs, of which cost control and scale-up production are the two cores.

Author Contributions: Conceptualization, C.-H.C. and Y.Z.; writing—original draft preparation, Y.Z.; writing—review and editing, C.-H.C. All authors have read and agreed to the published version of the manuscript.

Funding: This project was also partially funded by the National Science Foundation (NSF) under grant No. ECCS-1707506 and Scalable Nanomanufacturing program under Grant No. CBET-1449383.

Acknowledgments: We like to thank Alvin Chang for his assistance in improving the writing.

Conflicts of Interest: The authors declare no conflict of interest.

References

1. Yaghi, O.M.; Li, G.; Li, H. Selective binding and removal of guests in a microporous metal-organic framework. *Nature* **1995**, *378*, 703–706. [CrossRef]
2. Dietzel, P.D.; Besikiotis, V.; Blom, R. Application of metal-organic frameworks with coordinatively unsaturated metal sites in storage and separation of methane and carbon dioxide. *J. Mater. Chem.* **2009**, *19*, 7362–7370. [CrossRef]
3. Murray, L.J.; Dincă, M.; Long, J.R. Hydrogen storage in metal-organic frameworks. *Chem. Soc. Rev.* **2009**, *38*, 1294–1314. [CrossRef] [PubMed]
4. Li, B.; Wen, H.-M.; Zhou, W.; Chen, B. Porous metal-organic frameworks for gas storage and separation: What, how, and why? *J. Phys. Chem. Lett.* **2014**, *5*, 3468–3479. [CrossRef] [PubMed]
5. Li, H.; Wang, K.; Sun, Y.; Lollar, C.T.; Li, J.; Zhou, H.-C. Recent advances in gas storage and separation using metal-organic frameworks. *Mater. Today* **2018**, *21*, 108–121. [CrossRef]
6. Vlasova, E.; Yakimov, S.; Naidenko, E.; Kudrik, E.; Makarov, S. Application of metal-organic frameworks for purification of vegetable oils. *Food Chem.* **2016**, *190*, 103–109. [CrossRef] [PubMed]
7. Mukherjee, S.; Desai, A.V.; Ghosh, S.K. Potential of metal-organic frameworks for adsorptive separation of industrially and environmentally relevant liquid mixtures. *Coord. Chem. Rev.* **2018**, *367*, 82–126. [CrossRef]
8. Dhaka, S.; Kumar, R.; Deep, A.; Kurade, M.B.; Ji, S.-W.; Jeon, B.-H. Metal-organic frameworks (MOFs) for the removal of emerging contaminants from aquatic environments. *Coord. Chem. Rev.* **2019**, *380*, 330–352. [CrossRef]
9. Wang, H.; Zhao, S.; Liu, Y.; Yao, R.; Wang, X.; Cao, Y.; Ma, D.; Zou, M.; Cao, A.; Feng, X. Membrane adsorbers with ultrahigh metal-organic framework loading for high flux separations. *Nat. Commun.* **2019**, *10*, 1–9. [CrossRef]
10. Dhakshinamoorthy, A.; Li, Z.; Garcia, H. Catalysis and photocatalysis by metal organic frameworks. *Chem. Soc. Rev.* **2018**, *47*, 8134–8172. [CrossRef]
11. Kousik, S.; Velmathi, S. Engineering metal-organic framework catalysts for C–C and C–X coupling reactions: Advances in reticular approaches from 2014–2018. *Chem. Eur. J.* **2019**, *25*, 16451–16505. [CrossRef] [PubMed]
12. Pascanu, V.; González Miera, G.; Inge, A.K.; Martín-Matute, B.n. Metal-organic frameworks as catalysts for organic synthesis: A critical perspective. *J. Am. Chem. Soc.* **2019**, *141*, 7223–7234. [CrossRef] [PubMed]
13. Yang, D.; Gates, B.C. Catalysis by metal organic frameworks: Perspective and suggestions for future research. *ACS Catal.* **2019**, *9*, 1779–1798. [CrossRef]
14. Kreno, L.E.; Leong, K.; Farha, O.K.; Allendorf, M.; Van Duyne, R.P.; Hupp, J.T. Metal-organic framework materials as chemical sensors. *Chem. Rev.* **2012**, *112*, 1105–1125. [CrossRef] [PubMed]
15. Kumar, P.; Deep, A.; Kim, K.-H. Metal organic frameworks for sensing applications. *Trends Anal. Chem.* **2015**, *73*, 39–53. [CrossRef]
16. Chong, X.; Kim, K.-j.; Zhang, Y.; Li, E.; Ohodnicki, P.R.; Chang, C.-H.; Wang, A.X. Plasmonic nanopatch array with integrated metal-organic framework for enhanced infrared absorption gas sensing. *Nanotechnology* **2017**, *28*, 26LT01. [CrossRef]

17. Fang, X.; Zong, B.; Mao, S. Metal-organic framework-based sensors for environmental contaminant sensing. *Nano-Micro Lett.* **2018**, *10*, 64.
18. Chocarro-Ruiz, B.; Pérez-Carvajal, J.; Avci, C.; Calvo-Lozano, O.; Alonso, M.I.; Maspoch, D.; Lechuga, L.M. A CO_2 optical sensor based on self-assembled metal-organic framework nanoparticles. *J. Mater. Chem. A* **2018**, *6*, 13171–13177. [CrossRef]
19. Li, S.-L.; Xu, Q. Metal-organic frameworks as platforms for clean energy. *Energy Environ. Sci.* **2013**, *6*, 1656–1683. [CrossRef]
20. Bon, V. Metal-organic frameworks for energy-related applications. *Curr. Opin. Green Sustain. Chem.* **2017**, *4*, 44–49. [CrossRef]
21. Wu, H.B.; Lou, X.W.D. Metal-organic frameworks and their derived materials for electrochemical energy storage and conversion: Promises and challenges. *Sci. Adv.* **2017**, *3*, 9252.
22. Qiu, T.; Liang, Z.; Guo, W.; Tabassum, H.; Gao, S.; Zou, R. Metal-organic framework-based materials for energy conversion and storage. *ACS Energy Lett.* **2020**. [CrossRef]
23. Ló, Y.; Zhan, W.; He, Y.; Wang, Y.; Kong, X.; Kuang, Q.; Xie, Z.; Zheng, L. MOF-templated synthesis of porous Co_3O_4 concave nanocubes with high specific surface area and their gas sensing properties. *ACS Appl. Mater. Interfaces* **2014**, *6*, 4186–4195. [CrossRef] [PubMed]
24. Robson, R.; Abrahams, B.F.; Batten, S.R.; Gable, R.W.; Hoskins, B.F.; Liu, J. Crystal Engineering of Novel Materials Composed of Infinite Two-and Three-Dimensional Frameworks; ACS Symp. Ser. Am. Chem. Soc. **1992**, *499*, 256–273. [CrossRef]
25. Fujita, M.; Kwon, Y.J.; Washizu, S.; Ogura, K. Preparation, clathration ability, and catalysis of a two-dimensional square network material composed of cadmium (II) and 4,4'-bipyridine. *J. Am. Chem. Soc.* **1994**, *116*, 1151–1152.
26. Chui, S.S.-Y.; Lo, S.M.-F.; Charmant, J.P.; Orpen, A.G.; Williams, I.D. A chemically functionalizable nanoporous material [$Cu_3(TMA)_2(H_2O)_3$]$_n$. *Science* **1999**, *283*, 1148–1150. [PubMed]
27. Kim, K.-J.; Li, Y.J.; Kreider, P.B.; Chang, C.-H.; Wannenmacher, N.; Thallapally, P.K.; Ahn, H.-G. High-rate synthesis of Cu-BTC metal-organic frameworks. *Chem. Commun.* **2013**, *49*, 11518–11520.
28. Zhuang, J.; Kuo, C.-H.; Chou, L.-Y.; Liu, D.-Y.; Weerapana, E.; Tsung, C.-K. Optimized metal-organic-framework nanospheres for drug delivery: Evaluation of small-molecule encapsulation. *ACS Nano* **2014**, *8*, 2812–2819. [CrossRef]
29. Li, Y.-z.; Fu, Z.-h.; Xu, G. Metal-organic framework nanosheets: Preparation and applications. *Coord. Chem. Rev.* **2019**, *388*, 79–106. [CrossRef]
30. Park, K.H.; Kim, M.H.; Im, S.H.; Park, O.O. Electrically bistable Ag nanocrystal-embedded metal-organic framework microneedles. *RSC Adv.* **2016**, *6*, 64885–64889.
31. Hou, J.; Sapnik, A.F.; Bennett, T.D. Metal-organic framework gels and monoliths. *Chem. Sci.* **2020**. [CrossRef] [PubMed]
32. Zacher, D.; Shekhah, O.; Wöll, C.; Fischer, R.A. Thin films of metal-organic frameworks. *Chem. Soc. Rev.* **2009**, *38*, 1418–1429. [CrossRef] [PubMed]
33. Li, W. Metal-organic framework membranes: Production, modification, and applications. *Prog. Mater. Sci.* **2019**, *100*, 21–63. [CrossRef]
34. Li, S.; Limbach, R.; Longley, L.; Shirzadi, A.A.; Walmsley, J.C.; Johnstone, D.N.; Midgley, P.A.; Wondraczek, L.; Bennett, T.D. Mechanical properties and processing techniques of bulk metal-organic framework glasses. *J. Am. Chem. Soc.* **2018**, *141*, 1027–1034. [CrossRef]
35. Longley, L.; Collins, S.M.; Li, S.; Smales, G.J.; Erucar, I.; Qiao, A.; Hou, J.; Doherty, C.M.; Thornton, A.W.; Hill, A.J. Flux melting of metal-organic frameworks. *Chem. Sci.* **2019**, *10*, 3592–3601. [CrossRef]
36. Qiao, A.; Tao, H.; Carson, M.P.; Aldrich, S.W.; Thirion, L.M.; Bennett, T.D.; Mauro, J.C.; Yue, Y. Optical properties of a melt-quenched metal-organic framework glass. *Opt. Lett.* **2019**, *44*, 1623–1625. [CrossRef]
37. Liao, Z.; Xia, T.; Yu, E.; Cui, Y. Luminescent metal-organic framework thin films: from preparation to biomedical sensing applications. *Crystals* **2018**, *8*, 338. [CrossRef]
38. Cui, Y.; Zhang, J.; He, H.; Qian, G. Photonic functional metal-organic frameworks. *Chem. Soc. Rev.* **2018**, *47*, 5740–5785. [CrossRef]
39. Song, X.; Wang, X.; Li, Y.; Zheng, C.; Zhang, B.; Di, C.a.; Li, F.; Jin, C.; Mi, W.; Chen, L. 2D semiconducting metal-organic framework thin films for organic spin valves. *Angew. Chem. Int. Ed.* **2019**. [CrossRef]

40. De Luna, P.; Liang, W.; Mallick, A.; Shekhah, O.; García de Arquer, F.P.; Proppe, A.H.; Todorović, P.; Kelley, S.O.; Sargent, E.H.; Eddaoudi, M. Metal-organic framework thin films on high-curvature nanostructures toward tandem electrocatalysis. *ACS Appl. Mater. Interfaces* **2018**, *10*, 31225–31232. [CrossRef]
41. Kim, K.-J.; Chong, X.; Kreider, P.B.; Ma, G.; Ohodnicki, P.R.; Baltrus, J.P.; Wang, A.X.; Chang, C.-H. Plasmonics-enhanced metal-organic framework nanoporous films for highly sensitive near-infrared absorption. *J. Mater. Chem. C* **2015**, *3*, 2763–2767. [CrossRef]
42. Chong, X.; Kim, K.-J.; Li, E.; Zhang, Y.; Ohodnicki, P.R.; Chang, C.-H.; Wang, A.X. Near-infrared absorption gas sensing with metal-organic framework on optical fibers. *Sens. Actuators B Chem.* **2016**, *232*, 43–51. [CrossRef]
43. Chong, X.Y.; Zhang, Y.J.; Li, E.W.; Kim, K.J.; Ohodnicki, P.R.; Chang, C.H.; Wang, A.X. Surface-enhanced infrared absorption: Pushing the frontier for on-chip gas sensing. *ACS Sens.* **2018**, *3*, 230–238. [CrossRef] [PubMed]
44. Bai, W.; Li, S.; Ma, J.; Cao, W.; Zheng, J. Ultrathin 2D metal-organic framework (nanosheets and nanofilms)-based x D-2D hybrid nanostructures as biomimetic enzymes and supercapacitors. *J. Mater. Chem. A* **2019**, *7*, 9086–9098. [CrossRef]
45. Ahmad, S.; Liu, J.; Ji, W.; Sun, L. Metal-organic framework thin film-based dye sensitized solar cells with enhanced photocurrent. *Materials* **2018**, *11*, 1868. [CrossRef]
46. Luo, J.; Li, Y.; Zhang, H.; Wang, A.; Lo, W.S.; Dong, Q.; Wong, N.; Povinelli, C.; Shao, Y.; Chereddy, S. A metal-organic framework thin film for selective Mg^{2+} transport. *Angew. Chem. Int. Ed.* **2019**, *58*, 15313–15317. [CrossRef]
47. Koros, W.; Ma, Y.; Shimidzu, T. Terminology for membranes and membrane processes (IUPAC Recommendations 1996). *Pure Appl. Chem.* **1996**, *68*, 1479–1489. [CrossRef]
48. Ter Minassian-Saraga, L. Thin films including layers: Terminology in relation to their preparation and characterization (IUPAC Recommendations 1994). *Pure Appl. Chem.* **1994**, *66*, 1667–1738. [CrossRef]
49. Liu, T.-Y.; Yuan, H.-G.; Liu, Y.-Y.; Ren, D.; Su, Y.-C.; Wang, X. Metal-organic framework nanocomposite thin films with interfacial bindings and self-standing robustness for high water flux and enhanced ion selectivity. *ACS Nano* **2018**, *12*, 9253–9265. [CrossRef]
50. Venkatasubramanian, A.; Navaei, M.; Bagnall, K.R.; McCarley, K.C.; Nair, S.; Hesketh, P.J. Gas Adsorption characteristics of metal-organic frameworks via quartz crystal microbalance techniques. *J. Phys. Chem. C* **2012**, *116*, 15313–15321. [CrossRef]
51. Wannapaiboon, S.; Tu, M.; Sumida, K.; Khaletskaya, K.; Furukawa, S.; Kitagawa, S.; Fischer, R.A. Hierarchical structuring of metal-organic framework thin-films on quartz crystal microbalance (QCM) substrates for selective adsorption applications. *J. Mater. Chem. A* **2015**, *3*, 23385–23394. [CrossRef]
52. Guo, W.; Zha, M.; Wang, Z.; Redel, E.; Xu, Z.; Wöll, C. Improving the loading capacity of metal-organic framework thin films using optimized linkers. *ACS Appl. Mater. Interfaces* **2016**, *8*, 24699–24702. [CrossRef] [PubMed]
53. Shekhah, O. Layer-by-layer method for the synthesis and growth of surface mounted metal-organic frameworks (SURMOFs). *Materials* **2010**, *3*, 1302–1315. [CrossRef]
54. Mártire, A.P.; Segovia, G.M.; Azzaroni, O.; Rafti, M.; Marmisollé, W. Layer-by-layer integration of conducting polymers and metal organic frameworks onto electrode surfaces: Enhancement of the oxygen reduction reaction through electrocatalytic nanoarchitectonics. *Mol. Syst. Des. Eng.* **2019**, *4*, 893–900. [CrossRef]
55. Bhardwaj, S.K.; Bhardwaj, N.; Kaur, R.; Mehta, J.; Sharma, A.L.; Kim, K.-H.; Deep, A. An overview of different strategies to introduce conductivity in metal-organic frameworks and miscellaneous applications thereof. *J. Mater. Chem. A* **2018**, *6*, 14992–15009. [CrossRef]
56. Yoo, Y.; Jeong, H.-K. Rapid fabrication of metal organic framework thin films using microwave-induced thermal deposition. *Chem. Commun.* **2008**, *21*, 2441–2443. [CrossRef]
57. Kim, K.-J.; Zhang, Y.; Kreider, P.B.; Chong, X.; Wang, A.X.; Ohodnicki, P.R., Jr.; Baltrus, J.P.; Chang, C.-H. Nucleation and growth of oriented metal-organic framework thin films on thermal SiO_2 surface. *Thin Solid Films* **2018**, *659*, 24–35. [CrossRef]
58. Dimitrakakis, C.; Easton, C.D.; Muir, B.W.; Ladewig, B.P.; Hill, M.R. Spatial control of zeolitic imidazolate framework growth on flexible substrates. *Cryst. Growth Des.* **2013**, *13*, 4411–4417. [CrossRef]

59. Hermes, S.; Schröder, F.; Chelmowski, R.; Wöll, C.; Fischer, R.A. Selective nucleation and growth of metal-organic open framework thin films on patterned COOH/CF$_3$-terminated self-assembled monolayers on Au (111). *J. Am. Chem. Soc.* **2005**, *127*, 13744–13745. [CrossRef]
60. Scherb, C.; Williams, J.J.; Hinterholzinger, F.; Bauer, S.; Stock, N.; Bein, T. Implementing chemical functionality into oriented films of metal-organic frameworks on self-assembled monolayers. *J. Mater. Chem.* **2011**, *21*, 14849–14856. [CrossRef]
61. Julien, P.A.; Mottillo, C.; Friščić, T. Metal-organic frameworks meet scalable and sustainable synthesis. *Green Chem.* **2017**, *19*, 2729–2747. [CrossRef]
62. Ji, Y.; Qian, W.; Yu, Y.; An, Q.; Liu, L.; Zhou, Y.; Gao, C. Recent developments in nanofiltration membranes based on nanomaterials. *Chin. J. Chem. Eng.* **2017**, *25*, 1639–1652. [CrossRef]
63. Liu, Y.; Ban, Y.; Yang, W. Microstructural engineering and architectural design of metal-organic framework membranes. *Adv. Mater.* **2017**, *29*, 1606949. [CrossRef] [PubMed]
64. Echaide-Górriz, C.; Clément, C.; Cacho-Bailo, F.; Téllez, C.; Coronas, J. New strategies based on microfluidics for the synthesis of metal-organic frameworks and their membranes. *J. Mater. Chem. A* **2018**, *6*, 5485–5506. [CrossRef]
65. Zhu, J.; Hou, J.; Uliana, A.; Zhang, Y.; Tian, M.; Van der Bruggen, B. The rapid emergence of two-dimensional nanomaterials for high-performance separation membranes. *J. Mater. Chem. A* **2018**, *6*, 3773–3792. [CrossRef]
66. Shekhah, O.; Chernikova, V.; Belmabkhout, Y.; Eddaoudi, M. Metal-organic framework membranes: From fabrication to gas separation. *Crystals* **2018**, *8*, 412. [CrossRef]
67. Lin, R.; Hernandez, B.V.; Ge, L.; Zhu, Z. Metal organic framework based mixed matrix membranes: An overview on filler/polymer interfaces. *J. Mater. Chem. A* **2018**, *6*, 293–312. [CrossRef]
68. Jeazet, H.B.T.; Staudt, C.; Janiak, C. Metal-organic frameworks in mixed-matrix membranes for gas separation. *Dalton Trans.* **2012**, *41*, 14003–14027. [CrossRef]
69. Pettinari, C.; Marchetti, F.; Mosca, N.; Tosi, G.; Drozdov, A. Application of metal-organic frameworks. *Polym. Int.* **2017**, *66*, 731–744. [CrossRef]
70. Mahata, P.; Mondal, S.K.; Singha, D.K.; Majee, P. Luminescent rare-earth-based MOFs as optical sensors. *Dalton Trans.* **2017**, *46*, 301–328. [CrossRef]
71. Lustig, W.P.; Mukherjee, S.; Rudd, N.D.; Desai, A.V.; Li, J.; Ghosh, S.K. Metal-organic frameworks: Functional luminescent and photonic materials for sensing applications. *Chem. Soc. Rev.* **2017**, *46*, 3242–3285. [CrossRef] [PubMed]
72. Stassen, I.; Burtch, N.; Talin, A.; Falcaro, P.; Allendorf, M.; Ameloot, R. An updated roadmap for the integration of metal-organic frameworks with electronic devices and chemical sensors. *Chem. Soc. Rev.* **2017**, *46*, 3185–3241. [CrossRef] [PubMed]
73. Xu, K.; Fu, C.; Gao, Z.; Wei, F.; Ying, Y.; Xu, C.; Fu, G. Nanomaterial-based gas sensors: A review. *Instrum. Sci. Technol.* **2018**, *46*, 115–145. [CrossRef]
74. Li, Y.; Xiao, A.-S.; Zou, B.; Zhang, H.-X.; Yan, K.-L.; Lin, Y. Advances of metal-organic frameworks for gas sensing. *Polyhedron* **2018**, *154*, 83–97. [CrossRef]
75. Alrammouz, R.; Podlecki, J.; Abboud, P.; Sorli, B.; Habchi, R. A review on flexible gas sensors: From materials to devices. *Sens. Actuators A Phys.* **2018**, *284*, 209–231. [CrossRef]
76. Kuyuldar, S.; Genna, D.T.; Burda, C. On the potential for nanoscale metal-organic frameworks for energy applications. *J. Mater. Chem. A* **2019**, *7*, 21545–21576. [CrossRef]
77. Sosa, J.D.; Bennett, T.F.; Nelms, K.J.; Liu, B.M.; Tovar, R.C.; Liu, Y. Metal-organic framework hybrid materials and their applications. *Crystals* **2018**, *8*, 325. [CrossRef]
78. Wang, S.; McGuirk, C.M.; d'Aquino, A.; Mason, J.A.; Mirkin, C.A. Metal-organic framework nanoparticles. *Adv. Mater.* **2018**, *30*, 1800202. [CrossRef]
79. Zhu, Q.-L.; Xu, Q. Metal-organic framework composites. *Chem. Soc. Rev.* **2014**, *43*, 5468–5512. [CrossRef]
80. Ameloot, R.; Vermoortele, F.; Vanhove, W.; Roeffaers, M.B.; Sels, B.F.; De Vos, D.E. Interfacial synthesis of hollow metal-organic framework capsules demonstrating selective permeability. *Nat. Chem.* **2011**, *3*, 382–387. [CrossRef]
81. Li, J.; Wu, Q.; Wu, J. Synthesis of nanoparticles via solvothermal and hydrothermal methods. *Handb. Nanopart.* **2015**. [CrossRef]

82. Cui, X.-Y.; Gu, Z.-Y.; Jiang, D.-Q.; Li, Y.; Wang, H.-F.; Yan, X.-P. In situ hydrothermal growth of metal-organic framework 199 films on stainless steel fibers for solid-phase microextraction of gaseous benzene homologues. *Anal. Chem.* **2009**, *81*, 9771–9777. [CrossRef] [PubMed]
83. Sheberla, D.; Sun, L.; Blood-Forsythe, M.A.; Er, S.l.; Wade, C.R.; Brozek, C.K.; Aspuru-Guzik, A.n.; Dincă, M. High electrical conductivity in Ni_3(2,3,6,7,10,11-hexaiminotriphenylene)$_2$, a semiconducting metal-organic graphene analogue. *J. Am. Chem. Soc.* **2014**, *136*, 8859–8862. [CrossRef] [PubMed]
84. Campbell, J.; Tokay, B. Controlling the size and shape of Mg-MOF-74 crystals to optimise film synthesis on alumina substrates. *Microporous Mesoporous Mat.* **2017**, *251*, 190–199. [CrossRef]
85. Liu, Y.; Ng, Z.; Khan, E.A.; Jeong, H.-K.; Ching, C.-b.; Lai, Z. Synthesis of continuous MOF-5 membranes on porous α-alumina substrates. *Microporous Mesoporous Mat.* **2009**, *118*, 296–301. [CrossRef]
86. Yoon, S.M.; Park, J.H.; Grzybowski, B.A. Large-area, freestanding mof films of planar, curvilinear, or micropatterned topographies. *Angew. Chem. Int. Ed.* **2017**, *56*, 127–132. [CrossRef]
87. Bux, H.; Chmelik, C.; van Baten, J.M.; Krishna, R.; Caro, J. Novel MOF-membrane for molecular sieving predicted by IR-diffusion studies and molecular modeling. *Adv. Mater.* **2010**, *22*, 4741–4743. [CrossRef]
88. Liu, C.; Wu, Y.-n.; Morlay, C.; Gu, Y.; Gebremariam, B.; Yuan, X.; Li, F. General deposition of metal-organic frameworks on highly adaptive organic-inorganic hybrid electrospun fibrous substrates. *ACS Appl. Mater. Interfaces* **2016**, *8*, 2552–2561. [CrossRef]
89. Van Vleet, M.J.; Weng, T.; Li, X.; Schmidt, J. In Situ, time-resolved, and mechanistic studies of metal-organic framework nucleation and growth. *Chem. Rev.* **2018**, *118*, 3681–3721. [CrossRef]
90. Liu, J.; Wöll, C. Surface-supported metal-organic framework thin films: Fabrication methods, applications, and challenges. *Chem. Soc. Rev.* **2017**, *46*, 5730–5770. [CrossRef]
91. Brower, L.J.; Gentry, L.K.; Napier, A.L.; Anderson, M.E. Tailoring the nanoscale morphology of HKUST-1 thin films via codeposition and seeded growth. *Beilstein J. Nanotechnol.* **2017**, *8*, 2307–2314. [CrossRef] [PubMed]
92. Bradshaw, D.; Garai, A.; Huo, J. Metal-organic framework growth at functional interfaces: Thin films and composites for diverse applications. *Chem. Soc. Rev.* **2012**, *41*, 2344–2381. [CrossRef] [PubMed]
93. Ulman, A. Formation and structure of self-assembled monolayers. *Chem. Rev.* **1996**, *96*, 1533–1554. [CrossRef]
94. Biemmi, E.; Scherb, C.; Bein, T. Oriented growth of the metal organic framework $Cu_3(BTC)_2(H_2O)_3 \cdot xH_2O$ tunable with functionalized self-assembled monolayers. *J. Am. Chem. Soc.* **2007**, *129*, 8054–8055. [CrossRef]
95. Liu, J.; Shekhah, O.; Stammer, X.; Arslan, H.K.; Liu, B.; Schüpbach, B.; Terfort, A.; Wöll, C. Deposition of metal-organic frameworks by liquid-phase epitaxy: The influence of substrate functional group density on film orientation. *Materials* **2012**, *5*, 1581–1592. [CrossRef]
96. Zacher, D.; Baunemann, A.; Hermes, S.; Fischer, R.A. Deposition of microcrystalline [$Cu_3(btc)_2$] and [$Zn_2(bdc)_2(dabco)$] at alumina and silica surfaces modified with patterned self assembled organic monolayers: Evidence of surface selective and oriented growth. *J. Mater. Chem.* **2007**, *17*, 2785–2792. [CrossRef]
97. Hinterholzinger, F.; Scherb, C.; Ahnfeldt, T.; Stock, N.; Bein, T. Oriented growth of the functionalized metal-organic framework CAU-1 ON–OH-and–COOH-terminated self-assembled monolayers. *Phys. Chem. Chem. Phys.* **2010**, *12*, 4515–4520. [CrossRef]
98. McCarthy, M.C.; Varela-Guerrero, V.; Barnett, G.V.; Jeong, H.-K. Synthesis of zeolitic imidazolate framework films and membranes with controlled microstructures. *Langmuir* **2010**, *26*, 14636–14641. [CrossRef]
99. Zhou, M.; Li, J.; Zhang, M.; Wang, H.; Lan, Y.; Wu, Y.-n.; Li, F.; Li, G. A polydopamine layer as the nucleation center of MOF deposition on "inert" polymer surfaces to fabricate hierarchically structured porous films. *Chem. Commun.* **2015**, *51*, 2706–2709. [CrossRef]
100. Bux, H.; Feldhoff, A.; Cravillon, J.; Wiebcke, M.; Li, Y.-S.; Caro, J. Oriented zeolitic imidazolate framework-8 membrane with sharp H_2/C_3H_8 molecular sieve separation. *Chem. Mater.* **2011**, *23*, 2262–2269. [CrossRef]
101. Papporello, R.L.; Miró, E.E.; Zamaro, J.M. Secondary growth of ZIF-8 films onto copper-based foils. Insight into surface interactions. *Microporous Mesoporous Mat.* **2015**, *211*, 64–72. [CrossRef]
102. Sun, Y.; Zhang, R.; Zhao, C.; Wang, N.; Xie, Y.; Li, J.-R. Self-modified fabrication of inner skin ZIF-8 tubular membranes by a counter diffusion assisted secondary growth method. *RSC Adv.* **2014**, *4*, 33007–33012. [CrossRef]
103. Abdollahian, Y.; Hauser, J.L.; Colinas, I.R.; Agustin, C.; Ichimura, A.S.; Oliver, S.R. IRMOF thin films templated by oriented zinc oxide nanowires. *Cryst. Growth Des.* **2014**, *14*, 1506–1509. [CrossRef]

104. Zhao, J.; Gong, B.; Nunn, W.T.; Lemaire, P.C.; Stevens, E.C.; Sidi, F.I.; Williams, P.S.; Oldham, C.J.; Walls, H.J.; Shepherd, S.D. Conformal and highly adsorptive metal-organic framework thin films via layer-by-layer growth on ALD-coated fiber mats. *J. Mater. Chem. A* **2015**, *3*, 1458–1464. [CrossRef]

105. Rivero Fuente, P.J.; Goicoechea Fernández, J.; Arregui San Martín, F.J. Layer-by-layer nano-assembly: A powerful tool for optical fiber sensing applications. *Sensors* **2019**, *19*, 683. [CrossRef] [PubMed]

106. Xiao, F.-X.; Pagliaro, M.; Xu, Y.-J.; Liu, B. Layer-by-layer assembly of versatile nanoarchitectures with diverse dimensionality: A new perspective for rational construction of multilayer assemblies. *Chem. Soc. Rev.* **2016**, *45*, 3088–3121. [CrossRef]

107. Shekhah, O.; Wang, H.; Kowarik, S.; Schreiber, F.; Paulus, M.; Tolan, M.; Sternemann, C.; Evers, F.; Zacher, D.; Fischer, R.A. Step-by-step route for the synthesis of metal-organic frameworks. *J. Am. Chem. Soc.* **2007**, *129*, 15118–15119. [CrossRef]

108. Wang, Z.; Wöll, C. Fabrication of metal-organic framework thin films using programmed layer-by-layer assembly techniques. *Adv. Mater. Technol.* **2019**, *4*, 1800413. [CrossRef]

109. Shekhah, O.; Fu, L.; Sougrat, R.; Belmabkhout, Y.; Cairns, A.J.; Giannelis, E.P.; Eddaoudi, M. Successful implementation of the stepwise layer-by-layer growth of MOF thin films on confined surfaces: Mesoporous silica foam as a first case study. *Chem. Commun.* **2012**, *48*, 11434–11436. [CrossRef]

110. Yao, M.S.; Lv, X.J.; Fu, Z.H.; Li, W.H.; Deng, W.H.; Wu, G.D.; Xu, G. Layer-by-layer assembled conductive metal-organic framework nanofilms for room-temperature chemiresistive sensing. *Angew. Chem. Int. Ed.* **2017**, *56*, 16510–16514. [CrossRef]

111. Shekhah, O.; Liu, J.; Fischer, R.; Wöll, C. MOF thin films: Existing and future applications. *Chem. Soc. Rev.* **2011**, *40*, 1081–1106. [CrossRef] [PubMed]

112. Zhuang, J.-L.; Terfort, A.; Wöll, C. Formation of oriented and patterned films of metal-organic frameworks by liquid phase epitaxy: A review. *Coord. Chem. Rev.* **2016**, *307*, 391–424. [CrossRef]

113. Chernikova, V.; Shekhah, O.; Spanopoulos, I.; Trikalitis, P.N.; Eddaoudi, M. Liquid phase epitaxial growth of heterostructured hierarchical MOF thin films. *Chem. Commun.* **2017**, *53*, 6191–6194. [CrossRef]

114. Stavila, V.; Volponi, J.; Katzenmeyer, A.M.; Dixon, M.C.; Allendorf, M.D. Kinetics and mechanism of metal-organic framework thin film growth: Systematic investigation of HKUST-1 deposition on QCM electrodes. *Chem. Sci.* **2012**, *3*, 1531–1540. [CrossRef]

115. Shekhah, O.; Hirai, K.; Wang, H.; Uehara, H.; Kondo, M.; Diring, S.; Zacher, D.; Fischer, R.A.; Sakata, O.; Kitagawa, S. MOF-on-MOF heteroepitaxy: Perfectly oriented [$Zn_2(ndc)_2(dabco)$]$_n$ grown on [$Cu_2(ndc)_2(dabco)$]$_n$ thin films. *Dalton Trans.* **2011**, *40*, 4954–4958. [CrossRef]

116. Heinke, L.; Cakici, M.; Dommaschk, M.; Grosjean, S.; Herges, R.; Bräse, S.; Wöll, C. Photoswitching in two-component surface-mounted metal-organic frameworks: Optically triggered release from a molecular container. *ACS Nano* **2014**, *8*, 1463–1467. [CrossRef]

117. Liu, B.; Tu, M.; Zacher, D.; Fischer, R.A. Multi variant surface mounted metal-organic frameworks. *Adv. Funct. Mater.* **2013**, *23*, 3790–3798. [CrossRef]

118. Li, W.-J.; Tu, M.; Cao, R.; Fischer, R.A. Metal-organic framework thin films: Electrochemical fabrication techniques and corresponding applications & perspectives. *J. Mater. Chem. A* **2016**, *4*, 12356–12369.

119. Jiang, D.; Burrows, A.D.; Xiong, Y.; Edler, K.J. Facile synthesis of crack-free metal-organic framework films on alumina by a dip-coating route in the presence of polyethylenimine. *J. Mater. Chem. A* **2013**, *1*, 5497–5500. [CrossRef]

120. Chaudhari, A.K.; Han, I.; Tan, J.C. Multifunctional supramolecular hybrid materials constructed from hierarchical self-ordering of in situ generated metal-organic framework (MOF) nanoparticles. *Adv. Mater.* **2015**, *27*, 4438–4446. [CrossRef]

121. Huang, Y.; Tao, C.-a.; Chen, R.; Sheng, L.; Wang, J. Comparison of fabrication methods of metal-organic framework optical thin films. *Nanomaterials* **2018**, *8*, 676. [CrossRef] [PubMed]

122. Horcajada, P.; Serre, C.; Grosso, D.; Boissiere, C.; Perruchas, S.; Sanchez, C.; Férey, G. Colloidal route for preparing optical thin films of nanoporous metal-organic frameworks. *Adv. Mater.* **2009**, *21*, 1931–1935. [CrossRef]

123. Lu, G.; Hupp, J.T. Metal-organic frameworks as sensors: A ZIF-8 based Fabry-Pérot device as a selective sensor for chemical vapors and gases. *J. Am. Chem. Soc.* **2010**, *132*, 7832–7833. [CrossRef]

124. Eslava, S.; Zhang, L.; Esconjauregui, S.; Yang, J.; Vanstreels, K.; Baklanov, M.R.; Saiz, E. Metal-organic framework ZIF-8 films as low-κ dielectrics in microelectronics. *Chem. Mater.* **2012**, *25*, 27–33. [CrossRef]

125. Chernikova, V.; Shekhah, O.; Eddaoudi, M. Advanced fabrication method for the preparation of MOF thin films: Liquid-phase epitaxy approach meets spin coating method. *ACS Appl. Mater. Interfaces* **2016**, *8*, 20459–20464. [CrossRef]
126. Burmann, P.; Zornoza, B.; Téllez, C.; Coronas, J. Mixed matrix membranes comprising MOFs and porous silicate fillers prepared via spin coating for gas separation. *Chem. Eng. Sci.* **2014**, *107*, 66–75. [CrossRef]
127. Hoseini, S.J.; Bahrami, M.; Nabavizadeh, S.M. ZIF-8 nanoparticles thin film at an oil-water interface as an electrocatalyst for the methanol oxidation reaction without the application of noble metals. *New J. Chem.* **2019**, *43*, 15811–15822. [CrossRef]
128. Li, L.; Jiao, X.; Chen, D.; Li, C. One-step asymmetric growth of continuous metal-organic framework thin films on two-dimensional colloidal crystal arrays: A facile approach toward multifunctional superstructures. *Cryst. Growth Des.* **2016**, *16*, 2700–2707. [CrossRef]
129. Szelagowska-Kunstman, K.; Cyganik, P.; Goryl, M.; Zacher, D.; Puterova, Z.; Fischer, R.A.; Szymonski, M. Surface structure of metal-organic framework grown on self-assembled monolayers revealed by high-resolution atomic force microscopy. *J. Am. Chem. Soc.* **2008**, *130*, 14446–14447. [CrossRef]
130. Katayama, Y.; Kalaj, M.; Barcus, K.S.; Cohen, S.M. Self-assembly of metal-organic framework (MOF) nanoparticle monolayers and free-standing multilayers. *J. Am. Chem. Soc.* **2019**, *141*, 20000–20003. [CrossRef]
131. Kwon, H.T.; Jeong, H.-K. In situ synthesis of thin zeolitic-imidazolate framework ZIF-8 membranes exhibiting exceptionally high propylene/propane separation. *J. Am. Chem. Soc.* **2013**, *135*, 10763–10768. [CrossRef] [PubMed]
132. Kwon, H.T.; Jeong, H.-K. Improving propylene/propane separation performance of Zeolitic-Imidazolate framework ZIF-8 Membranes. *Chem. Eng. Sci.* **2015**, *124*, 20–26. [CrossRef]
133. Yao, J.; Dong, D.; Li, D.; He, L.; Xu, G.; Wang, H. Contra-diffusion synthesis of ZIF-8 films on a polymer substrate. *Chem. Commun.* **2011**, *47*, 2559–2561. [CrossRef]
134. Shamsaei, E.; Lin, X.; Low, Z.-X.; Abbasi, Z.; Hu, Y.; Liu, J.Z.; Wang, H. Aqueous phase synthesis of ZIF-8 membrane with controllable location on an asymmetrically porous polymer substrate. *ACS Appl. Mater. Interfaces* **2016**, *8*, 6236–6244. [CrossRef]
135. Barankova, E.; Tan, X.; Villalobos, L.F.; Litwiller, E.; Peinemann, K.V. A metal chelating porous polymeric support: The missing link for a defect-free metal-organic framework composite membrane. *Angew. Chem. Int. Ed.* **2017**, *56*, 2965–2968. [CrossRef]
136. Schoedel, A.; Scherb, C.; Bein, T. Oriented nanoscale films of metal-organic frameworks by room-temperature gel-layer synthesis. *Angew. Chem. Int. Ed.* **2010**, *49*, 7225–7228. [CrossRef]
137. Ameloot, R.; Gobechiya, E.; Uji-i, H.; Martens, J.A.; Hofkens, J.; Alaerts, L.; Sels, B.F.; De Vos, D.E. Direct patterning of oriented metal-organic framework crystals via control over crystallization kinetics in clear precursor solutions. *Adv. Mater.* **2010**, *22*, 2685–2688. [CrossRef]
138. Zhuang, J.L.; Ceglarek, D.; Pethuraj, S.; Terfort, A. Rapid room-temperature synthesis of metal-organic framework HKUST-1 crystals in bulk and as oriented and patterned thin films. *Adv. Funct. Mater.* **2011**, *21*, 1442–1447. [CrossRef]
139. Zhuang, J.L.; Ar, D.; Yu, X.J.; Liu, J.X.; Terfort, A. Patterned deposition of metal-organic frameworks onto plastic, paper, and textile substrates by inkjet printing of a precursor solution. *Adv. Mater.* **2013**, *25*, 4631–4635. [CrossRef]
140. Bowser, B.H.; Brower, L.J.; Ohnsorg, M.L.; Gentry, L.K.; Beaudoin, C.K.; Anderson, M.E. Comparison of surface-bound and free-standing variations of HKUST-1 MOFs: Effect of activation and ammonia exposure on morphology, crystallinity, and composition. *Nanomaterials* **2018**, *8*, 650. [CrossRef]
141. Melgar, V.M.A.; Kwon, H.T.; Kim, J. Direct spraying approach for synthesis of ZIF-7 membranes by electrospray deposition. *J. Membr. Sci.* **2014**, *459*, 190–196. [CrossRef]
142. Xiao, Y.; Guo, X.; Huang, H.; Yang, Q.; Huang, A.; Zhong, C. Synthesis of MIL-88B (Fe)/Matrimid mixed-matrix membranes with high hydrogen permselectivity. *RSC Adv.* **2015**, *5*, 7253–7259. [CrossRef]
143. Al-Kutubi, H.; Gascon, J.; Sudhölter, E.J.; Rassaei, L. Electrosynthesis of metal-organic frameworks: Challenges and opportunities. *ChemElectroChem* **2015**, *2*, 462–474. [CrossRef]
144. Ameloot, R.; Stappers, L.; Fransaer, J.; Alaerts, L.; Sels, B.F.; De Vos, D.E. Patterned growth of metal-organic framework coatings by electrochemical synthesis. *Chem. Mater.* **2009**, *21*, 2580–2582. [CrossRef]

145. Li, W.-J.; Feng, J.-F.; Lin, Z.-J.; Yang, Y.-L.; Yang, Y.; Wang, X.-S.; Gao, S.-Y.; Cao, R. Patterned growth of luminescent metal-organic framework films: A versatile electrochemically-assisted microwave deposition method. *Chem. Commun.* **2016**, *52*, 3951–3954. [CrossRef]
146. Campagnol, N.; Van Assche, T.; Boudewijns, T.; Denayer, J.; Binnemans, K.; De Vos, D.; Fransaer, J. High pressure, high temperature electrochemical synthesis of metal-organic frameworks: Films of MIL-100 (Fe) and HKUST-1 in different morphologies. *J. Mater. Chem. A* **2013**, *1*, 5827–5830. [CrossRef]
147. Li, M.; Dincă, M. Reductive electrosynthesis of crystalline metal-organic frameworks. *J. Am. Chem. Soc.* **2011**, *133*, 12926–12929. [CrossRef]
148. Li, M.; Dincă, M. Selective formation of biphasic thin films of metal-organic frameworks by potential-controlled cathodic electrodeposition. *Chem. Sci.* **2014**, *5*, 107–111. [CrossRef]
149. Hod, I.; Bury, W.; Karlin, D.M.; Deria, P.; Kung, C.W.; Katz, M.J.; So, M.; Klahr, B.; Jin, D.; Chung, Y.W. Directed growth of electroactive metal-organic framework thin films using electrophoretic deposition. *Adv. Mater.* **2014**, *26*, 6295–6300. [CrossRef]
150. Zhu, H.; Liu, H.; Zhitomirsky, I.; Zhu, S. Preparation of metal-organic framework films by electrophoretic deposition method. *Mater. Lett.* **2015**, *142*, 19–22. [CrossRef]
151. Martinez Joaristi, A.; Juan-Alcañiz, J.; Serra-Crespo, P.; Kapteijn, F.; Gascon, J. Electrochemical synthesis of some archetypical Zn^{2+}, Cu^{2+}, and Al^{3+} metal organic frameworks. *Cryst. Growth Des.* **2012**, *12*, 3489–3498. [CrossRef]
152. Hauser, J.L.; Tso, M.; Fitchmun, K.; Oliver, S.R. Anodic electrodeposition of several metal organic framework thin films on indium tin oxide glass. *Cryst. Growth Des.* **2019**, *19*, 2358–2365. [CrossRef]
153. Alizadeh, S.; Nematollahi, D. Convergent and divergent paired electrodeposition of metal-organic framework thin films. *Sci. Rep.* **2019**, *9*, 1–13. [CrossRef] [PubMed]
154. Campagnol, N.; Van Assche, T.R.; Li, M.; Stappers, L.; Dincă, M.; Denayer, J.F.; Binnemans, K.; De Vos, D.E.; Fransaer, J. On the electrochemical deposition of metal-organic frameworks. *J. Mater. Chem. A* **2016**, *4*, 3914–3925. [CrossRef]
155. Li, J.; Cao, W.; Mao, Y.; Ying, Y.; Sun, L.; Peng, X. Zinc hydroxide nanostrands: Unique precursors for synthesis of ZIF-8 thin membranes exhibiting high size-sieving ability for gas separation. *CrystEngComm* **2014**, *16*, 9788–9791. [CrossRef]
156. Zou, X.; Zhu, G.; Hewitt, I.J.; Sun, F.; Qiu, S. Synthesis of a metal-organic framework film by direct conversion technique for VOCs sensing. *Dalton Trans.* **2009**, 3009–3013. [CrossRef]
157. Abuzalat, O.; Wong, D.; Elsayed, M.; Park, S.; Kim, S. Sonochemical fabrication of Cu(II) and Zn(II) metal-organic framework films on metal substrates. *Ultrason. Sonochem.* **2018**, *45*, 180–188. [CrossRef]
158. Kang, Z.; Xue, M.; Fan, L.; Ding, J.; Guo, L.; Gao, L.; Qiu, S. "Single nickel source" in situ fabrication of a stable homochiral MOF membrane with chiral resolution properties. *Chem. Commun.* **2013**, *49*, 10569–10571. [CrossRef]
159. Zhan, W.-w.; Kuang, Q.; Zhou, J.-z.; Kong, X.-j.; Xie, Z.-x.; Zheng, L.-s. Semiconductor@metal-organic framework core-shell heterostructures: A case of ZnO@ZIF-8 nanorods with selective photoelectrochemical response. *J. Am. Chem. Soc.* **2013**, *135*, 1926–1933. [CrossRef]
160. Khaletskaya, K.; Turner, S.; Tu, M.; Wannapaiboon, S.; Schneemann, A.; Meyer, R.; Ludwig, A.; Van Tendeloo, G.; Fischer, R.A. Self-directed localization of ZIF-8 thin film formation by conversion of ZnO nanolayers. *Adv. Funct. Mater.* **2014**, *24*, 4804–4811. [CrossRef]
161. Mao, Y.; Cao, W.; Li, J.; Liu, Y.; Ying, Y.; Sun, L.; Peng, X. Enhanced gas separation through well-intergrown MOF membranes: Seed morphology and crystal growth effects. *J. Mater. Chem. A* **2013**, *1*, 11711–11716. [CrossRef]
162. Zhang, Y.; Gao, Q.; Lin, Z.; Zhang, T.; Xu, J.; Tan, Y.; Tian, W.; Jiang, L. Constructing free standing metal organic framework MIL-53 membrane based on anodized aluminum oxide precursor. *Sci. Rep.* **2014**, *4*, 4947. [CrossRef] [PubMed]
163. Schäfer, P.; van der Veen, M.A.; Domke, K.F. Unraveling a two-step oxidation mechanism in electrochemical Cu-MOF synthesis. *Chem. Commun.* **2016**, *52*, 4722–4725. [CrossRef] [PubMed]
164. Stassen, I.; Campagnol, N.; Fransaer, J.; Vereecken, P.; De Vos, D.; Ameloot, R. Solvent-free synthesis of supported ZIF-8 films and patterns through transformation of deposited zinc oxide precursors. *CrystEngComm* **2013**, *15*, 9308–9311. [CrossRef]

165. Chen, Y.; Li, S.; Pei, X.; Zhou, J.; Feng, X.; Zhang, S.; Cheng, Y.; Li, H.; Han, R.; Wang, B. A solvent-free hot-pressing method for preparing metal-organic-framework coatings. *Angew. Chem. Int. Ed.* **2016**, *55*, 3419–3423. [CrossRef]
166. Lausund, K.B.; Nilsen, O. All-gas-phase synthesis of UiO-66 through modulated atomic layer deposition. *Nat. Commun.* **2016**, *7*, 13578. [CrossRef]
167. Stassen, I.; Styles, M.; Grenci, G.; Van Gorp, H.; Vanderlinden, W.; De Feyter, S.; Falcaro, P.; De Vos, D.; Vereecken, P.; Ameloot, R. Chemical vapour deposition of zeolitic imidazolate framework thin films. *Nat. Mater.* **2016**, *15*, 304. [CrossRef]
168. Fischer, D.; von Mankowski, A.; Ranft, A.; Vasa, S.K.; Linser, R.; Mannhart, J.; Lotsch, B.V. ZIF-8 films prepared by femtosecond pulsed-laser deposition. *Chem. Mater.* **2017**, *29*, 5148–5155. [CrossRef]
169. Li, W.; Su, P.; Li, Z.; Xu, Z.; Wang, F.; Ou, H.; Zhang, J.; Zhang, G.; Zeng, E. Ultrathin metal-organic framework membrane production by gel-vapour deposition. *Nat. Commun.* **2017**, *8*, 406. [CrossRef]
170. Makiura, R.; Motoyama, S.; Umemura, Y.; Yamanaka, H.; Sakata, O.; Kitagawa, H. Surface nano-architecture of a metal-organic framework. *Nat. Mater.* **2010**, *9*, 565. [CrossRef]
171. Makiura, R.; Kitagawa, H. Porous porphyrin nanoarchitectures on surfaces. *Eur. J. Inorg. Chem.* **2010**, *2010*, 3715–3724. [CrossRef]
172. Motoyama, S.; Makiura, R.; Sakata, O.; Kitagawa, H. Highly crystalline nanofilm by layering of porphyrin metal-organic framework sheets. *J. Am. Chem. Soc.* **2011**, *133*, 5640–5643. [CrossRef] [PubMed]
173. Makiura, R.; Konovalov, O. Bottom-up assembly of ultrathin sub-micron size metal-organic framework sheets. *Dalton Trans.* **2013**, *42*, 15931–15936. [CrossRef] [PubMed]
174. Makiura, R.; Konovalov, O. Interfacial growth of large-area single-layer metal-organic framework nanosheets. *Sci. Rep.* **2013**, *3*, 2506. [CrossRef]
175. Rubio-Giménez, V.c.; Tatay, S.; Volatron, F.; Martínez-Casado, F.J.; Martí-Gastaldo, C.; Coronado, E. High-quality metal-organic framework ultrathin films for electronically active interfaces. *J. Am. Chem. Soc.* **2016**, *138*, 2576–2584. [CrossRef]
176. Dong, R.; Pfeffermann, M.; Liang, H.; Zheng, Z.; Zhu, X.; Zhang, J.; Feng, X. Large-area, free-standing, two-dimensional supramolecular polymer single-layer sheets for highly efficient electrocatalytic hydrogen evolution. *Angew. Chem. Int. Ed.* **2015**, *54*, 12058–12063. [CrossRef]
177. Benito, J.; Sorribas, S.; Lucas, I.; Coronas, J.; Gascon, I. Langmuir-blodgett films of the metal-organic framework MIL-101 (Cr): Preparation, characterization, and CO_2 adsorption study using a QCM-based setup. *ACS Appl. Mater. Interfaces* **2016**, *8*, 16486–16492. [CrossRef]
178. Wang, Y.; Zhao, M.; Ping, J.; Chen, B.; Cao, X.; Huang, Y.; Tan, C.; Ma, Q.; Wu, S.; Yu, Y. Bioinspired design of ultrathin 2D bimetallic metal-organic-framework nanosheets used as biomimetic enzymes. *Adv. Mater.* **2016**, *28*, 4149–4155. [CrossRef]
179. Xu, G.; Yamada, T.; Otsubo, K.; Sakaida, S.; Kitagawa, H. Facile "modular assembly" for fast construction of a highly oriented crystalline MOF nanofilm. *J. Am. Chem. Soc.* **2012**, *134*, 16524–16527. [CrossRef]
180. Zhang, C.; Xiao, Y.; Liu, D.; Yang, Q.; Zhong, C. A hybrid zeolitic imidazolate framework membrane by mixed-linker synthesis for efficient CO_2 capture. *Chem. Commun.* **2013**, *49*, 600–602. [CrossRef]
181. Wu, B.; Lin, X.; Ge, L.; Wu, L.; Xu, T. A novel route for preparing highly proton conductive membrane materials with metal-organic frameworks. *Chem. Commun.* **2013**, *49*, 143–145. [CrossRef] [PubMed]
182. Mao, Y.; Li, J.; Cao, W.; Ying, Y.; Hu, P.; Liu, Y.; Sun, L.; Wang, H.; Jin, C.; Peng, X. General incorporation of diverse components inside metal-organic framework thin films at room temperature. *Nat. Commun.* **2014**, *5*, 5532. [CrossRef]
183. Gu, Z.-G.; Chen, Z.; Fu, W.-Q.; Wang, F.; Zhang, J. Liquid-phase epitaxy effective encapsulation of lanthanide coordination compounds into MOF film with homogeneous and tunable white-light emission. *ACS Appl. Mater. Interfaces* **2015**, *7*, 28585–28590. [CrossRef] [PubMed]
184. Fu, W.-Q.; Liu, M.; Gu, Z.-G.; Chen, S.-M.; Zhang, J. Liquid phase epitaxial growth and optical properties of photochromic guest-encapsulated MOF thin film. *Cryst. Growth Des.* **2016**, *16*, 5487–5492. [CrossRef]
185. Shekhah, O.; Arslan, H.K.; Chen, K.; Schmittel, M.; Maul, R.; Wenzel, W.; Wöll, C. Post-synthetic modification of epitaxially grown, highly oriented functionalized MOF thin films. *Chem. Commun.* **2011**, *47*, 11210–11212. [CrossRef]

186. Talin, A.A.; Centrone, A.; Ford, A.C.; Foster, M.E.; Stavila, V.; Haney, P.; Kinney, R.A.; Szalai, V.; El Gabaly, F.; Yoon, H.P. Tunable electrical conductivity in metal-organic framework thin-film devices. *Science* **2014**, *343*, 66–69. [CrossRef]
187. Hinterholzinger, F.M.; Wuttke, S.; Roy, P.; Preuße, T.; Schaate, A.; Behrens, P.; Godt, A.; Bein, T. Highly oriented surface-growth and covalent dye labeling of mesoporous metal-organic frameworks. *Dalton Trans.* **2012**, *41*, 3899–3901. [CrossRef]
188. Wang, Z.; Liu, J.; Arslan, H.K.; Grosjean, S.; Hagendorn, T.; Gliemann, H.; Bräse, S.; Wöll, C. Post-synthetic modification of metal-organic framework thin films using click chemistry: The importance of strained C–C triple bonds. *Langmuir* **2013**, *29*, 15958–15964. [CrossRef]
189. Tu, M.; Wannapaiboon, S.; Fischer, R.A. Programmed functionalization of SURMOFs via liquid phase heteroepitaxial growth and post-synthetic modification. *Dalton Trans.* **2013**, *42*, 16029–16035. [CrossRef]
190. Chen, Z.; Gu, Z.-G.; Fu, W.-Q.; Wang, F.; Zhang, J. A confined fabrication of perovskite quantum dots in oriented MOF thin film. *ACS Appl. Mater. Interfaces* **2016**, *8*, 28737–28742. [CrossRef]
191. Gassensmith, J.J.; Erne, P.M.; Paxton, W.F.; Valente, C.; Stoddart, J.F. Microcontact click printing for templating ultrathin films of metal-organic frameworks. *Langmuir* **2010**, *27*, 1341–1345. [CrossRef] [PubMed]
192. Liang, K.; Carbonell, C.; Styles, M.J.; Ricco, R.; Cui, J.; Richardson, J.J.; Maspoch, D.; Caruso, F.; Falcaro, P. Biomimetic replication of microscopic metal-organic framework patterns using printed protein patterns. *Adv. Mater.* **2015**, *27*, 7293–7298. [CrossRef] [PubMed]
193. Navarro, M.; Seoane, B.; Mateo, E.; Lahoz, R.; Germán, F.; Coronas, J. ZIF-8 micromembranes for gas separation prepared on laser-perforated brass supports. *J. Mater. Chem. A* **2014**, *2*, 11177–11184. [CrossRef]
194. Reboul, J.; Furukawa, S.; Horike, N.; Tsotsalas, M.; Hirai, K.; Uehara, H.; Kondo, M.; Louvain, N.; Sakata, O.; Kitagawa, S. Mesoscopic architectures of porous coordination polymers fabricated by pseudomorphic replication. *Nat. Mater.* **2012**, *11*, 717. [CrossRef]
195. Zhao, J.; Nunn, W.T.; Lemaire, P.C.; Lin, Y.; Dickey, M.D.; Oldham, C.J.; Walls, H.J.; Peterson, G.W.; Losego, M.D.; Parsons, G.N. Facile conversion of hydroxy double salts to metal-organic frameworks using metal oxide particles and atomic layer deposition thin-film templates. *J. Am. Chem. Soc.* **2015**, *137*, 13756–13759. [CrossRef]
196. Kaigala, G. V.; Lovchik, R. D.; Drechsler, U.; Delamarche, E. A vertical microfluidic probe. *Langmuir* **2011**, *27*, 5686–5693. [CrossRef]
197. Cui, J.; Gao, N.; Yin, X.; Zhang, W.; Liang, Y.; Tian, L.; Zhou, K.; Wang, S.; Li, G. Microfluidic synthesis of uniform single-crystalline MOF microcubes with a hierarchical porous structure. *Nanoscale* **2018**, *10*, 9192–9198. [CrossRef]
198. Surble, S.; Millange, F.; Serre, C.; Ferey, G.; Walton, R.I. An EXAFS study of the formation of a nanoporous metal-organic framework: Evidence for the retention of secondary building units during synthesis. *Chem. Commun.* **2006**, 1518–1520. [CrossRef]
199. Eddaoudi, M.; Moler, D.B.; Li, H.; Chen, B.; Reineke, T.M.; O'keeffe, M.; Yaghi, O.M. Modular chemistry: Secondary building units as a basis for the design of highly porous and robust metal-organic carboxylate frameworks. *Acc. Chem. Res.* **2001**, *34*, 319–330. [CrossRef]
200. Guillerm, V.; Gross, S.; Serre, C.; Devic, T.; Bauer, M.; Férey, G. A zirconium methacrylate oxocluster as precursor for the low-temperature synthesis of porous zirconium (IV) dicarboxylates. *Chem. Commun.* **2010**, *46*, 767–769. [CrossRef]

© 2020 by the authors. Licensee MDPI, Basel, Switzerland. This article is an open access article distributed under the terms and conditions of the Creative Commons Attribution (CC BY) license (http://creativecommons.org/licenses/by/4.0/).

Review

Plasmonic-Active Nanostructured Thin Films

Jay K. Bhattarai [1], Md Helal Uddin Maruf [2] and Keith J. Stine [1,*]

1. Department of Chemistry and Biochemistry, University of Missouri-St. Louis, Saint Louis, MO 63121, USA; jkbxv3@umsystem.edu
2. Department of Physics and Astronomy, University of Missouri–St. Louis, Saint Louis, MO 63121, USA; mm96f@mail.umsl.edu
* Correspondence: kstine@umsl.edu

Received: 31 December 2019; Accepted: 13 January 2020; Published: 16 January 2020

Abstract: Plasmonic-active nanomaterials are of high interest to scientists because of their expanding applications in the field for medicine and energy. Chemical and biological sensors based on plasmonic nanomaterials are well-established and commercially available, but the role of plasmonic nanomaterials on photothermal therapeutics, solar cells, super-resolution imaging, organic synthesis, etc. is still emerging. The effectiveness of the plasmonic materials on these technologies depends on their stability and sensitivity. Preparing plasmonics-active nanostructured thin films (PANTFs) on a solid substrate improves their physical stability. More importantly, the surface plasmons of thin film and that of nanostructures can couple in PANTFs enhancing the sensitivity. A PANTF can be used as a transducer for any of the three plasmonic-based sensing techniques, namely, the propagating surface plasmon, localized surface plasmon resonance, and surface-enhanced Raman spectroscopy-based sensing techniques. Additionally, continuous nanostructured metal films have an advantage for implementing electrical controls such as simultaneous sensing using both plasmonic and electrochemical techniques. Although research and development on PANTFs have been rapidly advancing, very few reviews on synthetic methods have been published. In this review, we provide some fundamental and practical aspects of plasmonics along with the recent advances in PANTFs synthesis, focusing on the advantages and shortcomings of the fabrication techniques. We also provide an overview of different types of PANTFs and their sensitivity for biosensing.

Keywords: plasmonics; localized surface plasmon resonance (LSPR); biosensing; thin film; gold nanostructures; lithography; nanohole array; nanofabrication

1. Introduction

Plasmonics is the study of interactions of light with plasmonic-active materials. It is an active field of research in nanotechnology bearing numerous potential applications in the field of biomedical science and energy, including chemical and biological sensing, photothermal therapeutics, super-resolution imaging, surface-enhanced Raman spectroscopy, solar cells, photocatalyst for organic synthesis, etc. [1–3]. The surface of plasmonic-active materials consists of confined surface plasmons, which are the coherent oscillation of free electrons [4]. A thick planar metal surface in contact with two dielectrics on the opposite boundaries support two independent surface plasmons. Interestingly, in a nanometer-thin planar metal-dielectric interface, surface plasmons on the opposite boundary couple with each other enhancing the electromagnetic field [5]. When the frequency of light waves (electromagnetic waves) matches that of surface plasmon, light waves can couple, excite, and propagate the surface plasmons at the metal-dielectric interface, called propagating surface plasmon resonance or surface plasmon polaritons (SPP) [5] (Figure 1a). However, simply hitting the smooth planar surface by light source does not match the frequencies of light and surface plasmon. The matching of frequencies can be achieved by enhancing the frequency of light using the attenuated total reflection or diffraction through different coupling devices [5]. The commonly used coupling devices for the excitation of surface plasmons on thin film are prism, waveguide, and grating. Among these, a prism in Kretschmann configuration has been extensively used in SPP-based biosensors as shown in Figure 1a. The excited surface plasmons can propagate tens to hundreds of microns along x and y axis on planar metal films, whereas the evanescent field produced on z axis decay exponentially with distance [5]. The change in refractive index near plasmonic-active planar thin film alters the evanescent waves changing the properties (angle, wavelength, intensity, and phase) of the light wave, which can be used as a probe for sensing molecules [6].

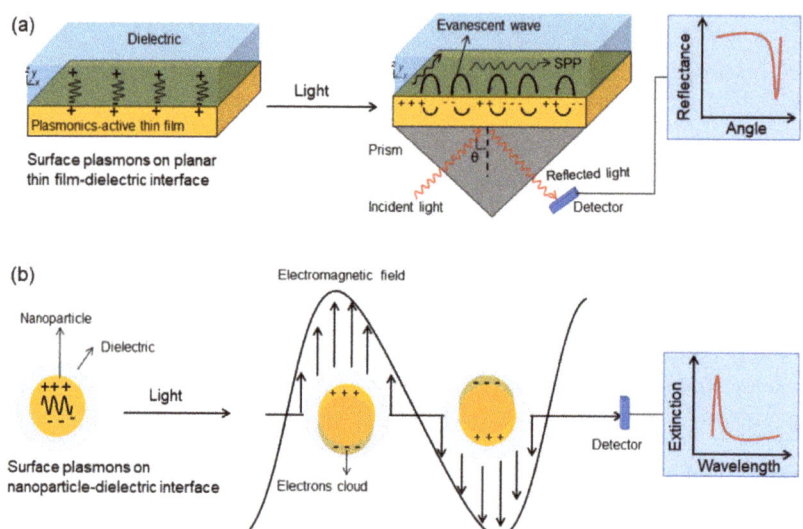

Figure 1. Schematic diagrams illustrating the excitation of surface plasmons from (**a**) thin film using prism as a coupling device, generating propagating surface plasmon or surface plasmon polariton (SPP), and (**b**) nanoparticle, generating a localized surface plasmon resonance (LSPR). The typical spectra that can be obtained after the surface plasmon excitation are shown on the right.

Unlike SPP, the surface plasmons of zero-dimensional nanostructures (e.g., nanoparticles, nanorods, nanostars, etc.) can be excited by the direct incident of light on nanostructures (Figure 1b). The excited surface plasmons resonate locally around the nanostructures with the frequency called localized surface plasmon resonance (LSPR) [4]. The LSPR depends on the shape, size, and composition of nanostructures as well as a change in the refractive index around nanostructures [7]. With the advancement of the plasmonics field, it is now possible to design a wide variety of zero-dimensional nanostructures with controlled shapes, sizes, and compositions [8]. This helps to tune the LSPR peak wavelength from UV to IR regions for desired applications. In general, pointy nanostructures have sharp peaks and are more sensitive toward change in the refractive index [9]. A smaller nanostructure of a particular material typically has sharper peak and initial peak wavelength falls at lower wavelength region compared to larger nanostructures of the same material [10]. However, the shift of peak wavelength with the change in refractive index is higher for larger nanoparticles but with a broader peak. The sensitivity of the silver nanomaterial is better than other metals, but it is prone to oxidation for use in many applications.

Although SPP-based biosensor has emerged as a leading technology for biosensing, LSPR-based biosensor bears great potential because of simple instrumentation and low cost with better sensitivity for molecular adsorption process [11]. Similar to SPP, LSPR also senses molecules and their interactions by monitoring the change in refractive index around nanostructures, commonly to the short-range (below 20 nm) owing to the shorter decay length of LSPR. However, recent studies have shown that even long-range refractive index sensing (>100 nm) is possible using ordered arrays and lines of plasmonic nanostructured films [12]. Therefore, the field of plasmonics is focusing on simultaneous excitation and coupling of surface plasmon of planar thin film and nanostructures, creating different nanostructured films or plasmonic metasurfaces. It was found that 30 nm thick continuous gold film under the array of disk or line nanostructures prepared using electron beam lithography cause plasmonic enhancement with the surface-enhanced Raman scattering (SERS) enhancement factor of 10^7 for thiophenol detection [13]. Plasmonic-active nanostructured thin films (PANTFs) are unique type of nanostructures, which have the properties of both thin films and individual nanoparticles. The necessity of exploring PANTFs arises due to (1) possibility of improving sensitivity of LSPR/SERS sensors due to coupling of neighboring nanostructures, (2) poor stability and reproducibility of individual nanostructures, (3) ease of systematic study from ordered structures, and (4) generating both LSPR and SPR simultaneously. The planar nanometer-thin film is 2D nanomaterial itself and is plasmonic-active, but we are not considering it as a PANTF. The PANTF must have other nanostructured features besides nanometer thickness and should be directly on contact with each other through the same materials throughout the substrate. Therefore, discrete nanostructures arrays, such as triangular nanoprism prepared using nanosphere lithography technique, are not the focus here as it is beyond the scope of this review. Excellent reviews on synthesis and applications of such structures have been covered elsewhere [14,15].

2. Composition of PANTFs

A wide variety of materials can produce plasmonic resonance. Some of the commonly explored materials for plasmonic-based applications are gold, silver, copper [16], aluminum [17], palladium [18], titanium nitride (TiN) [19], graphene [20], quantum dots [21], etc. Although Au and Ag are most explored for different applications, other material bears their unique advantages. The research on the plasmonic properties of other materials are still in the early phase. The Ag PANTFs have sharper LSPR peak and are highly sensitive compared to other material. However, quick oxidation of Ag can be a problem in many technologies. Researchers have deposited atomic layer thick alumina on top of Ag-based PANTFs by employing the atomic layer deposition technique, which avoids the oxidation of Ag without losing its plasmonic properties. Au PANTFs can trade-off the sensitivity of Ag PANTFs for stability.

Besides being inexpensive, aluminum is the material of choice when plasmonic peaks need to be in the UV region for applications. The plasmonic peak of gold and silver cannot reach UV region. The plasmonic peak of Al can also be tuned along the visible and near-infrared (NIR) regions. On the other hand, graphene-based PANTFs show plasmonic resonance peaks in the mid-IR region [20]. Additionally, it was found that fabricating Au nanostructures on graphene nanomesh greatly enhanced the LSPR peak [22]. TiN shows LSPR peaks on visible and NIR region with weaker plasmonic response than Au and Ag at room temperature but is very stable at higher temperature owing to its bulk melting point of 2930 °C [19]. This leads to the applications in the preparation of high-temperature nanophotonic devices.

3. Substrates

The PANTFs are most frequently prepared on SiO_2 substrates (traditional glass, fused silica and quartz), Si wafer, mica and different types of polymeric materials. The choice of the substrate depends on the objective of the study and available techniques. The SiO_2 substrates are the most widely used substrate for preparing PANTFs because of their low cost, easy availability, and transparency. These substrates can be easily coated with indium–tin–oxide (ITO) to make it conductive. The conductivity of the substrate is advantageous for the fabrication of PANTFs, such as during deposition by electrochemistry techniques [23] or to prevent charging during patterning on the electron beam lithography technique [24]. Although SiO_2 is an optimal substrate for transmission-based plasmonic sensing, Si wafer and mica can be used for reflection-based plasmonic sensing. Moreover, Si wafer-based

PANTFs prepared on the large area can be easily cut into smaller chips owing to its single crystalline nature. Among different polymeric substrate, poly (methyl methacrylate) (PMMA), polyethylene terephthalate (PET) and polycarbonate (PC) are commonly used for creating PANTFs. Besides being inexpensive and transparent, these substrates are highly flexible [25]. It was found that depositing thin layers of TiN on PMMA and PET shows similar plasmonic response as that of TiN deposited on SiO_2/Si [26]. In another study, large-area hexagonal gold nanohole arrays were fabricated by transferring gold film from silicon template to PC film by thermal annealing followed by the template-stripping method [27]. Chuo et al. took advantage of the flexibility of PET to prepare roll-to-roll embossing of plasmonic-active Au nanohole arrays on a 2000 ft production roll with the sensitivity of 180 nm/RIU [28]. This type of structure is advantageous for high-throughput chemicals and biomolecule detection.

4. Intermediate Layer for Stabilizing PANTFs

The intermediate layer between the substrate and PANTFs has a very important role in the stability and sensitivity of the structure. For many applications, PANTFs need to be strongly bound to the substrate. For example, in biosensing applications, PANTFs should be very stable in water or biological matrixes. Peeling off the PANTFs layer even in a small amount could be problematic for the reproduction of the data. Commonly, the glass substrates are coated with a thin layer of titanium or chromium before depositing the plasmonic materials. However, these non-plasmonic metal layers are known to dampen the sensitivity of the PANTFS due to absorption of light and interference for plasmon resonance, which broadens the LSPR peak [29,30]. They can also be the site for the non-specific binding of biomolecules [31]. Najiminaini et al. compared the effect of the Cr and Ti adhesion layer on the plasmonic property using a nanohole array [32]. They found that the optical resonance bandwidth of LSPR peak depends both on the composition and thickness of the adhesion layer. More importantly, removing 10 nm thick titanium adhesion layer from nanohole array by etching drastically decreases the optical resonance bandwidths. Another study has found that Cr can interdiffuse more with Au to form Cr–Au alloy than Ti, suggesting Ti is the optimal adhesion layer for Au-based PANTFs [33]. The adhesion layer has been engineered to minimize the plasmonic damping by depositing less than 1 nm adhesive, which prevents the layer to overlap with the hotspots of PANTFs [34]; however, thinner layer of adhesive also means low stability of PANTFs.

Different types of self-assembled monolayers (SAMs) with the terminal thio-group, including organosilanes, have been created to attach gold and silver-based PANTFs. Comparative measurements have shown that organic adhesion layers are better with the low plasmonic damping compared to metal-based adhesion layers [30]. However, organic adhesion layers are not compatible with many solvents used in the lift-off process during lithography techniques. A technology to stabilize PANTFs on the substrate without compensating the sensitivity is still a need.

5. Fabrication of Plasmonic-Active Nanostructured Thin Film

There are varieties of deposition methods available for the preparation of thin films, which can be broadly grouped as, (1) physical/chemical vapor deposition (2) sputtering (3) chemical/electrochemical deposition [35,36]. In vapor deposition techniques, a desired solid material is evaporated, commonly using elevated temperature, electricity, or electron beam, and deposited on the substrate as a thin film with or without reacting with substrate and hence called chemical vapor deposition (CVD) and physical vapor deposition (PVD), respectively. In sputtering techniques, the solid target is bombarded with high energy gaseous plasma or ions to eject atoms of target, which is deposited as a thin film on the substrate. Both vapor deposition and sputtering are commonly performed under vacuum to improve the quality of thin films. In chemical or electrochemical deposition method, precursor salts of the desired materials are reduced to deposit nanostructures on conductive film. Although these deposition techniques can deposit films of different thicknesses on the substrate by varying deposition parameters, they may not be able to create a wide variety of desired nanostructures on the substrate.

The PANTFs can be prepared on the substrate by employing different strategies as shown in Figure 2. The first and popular method is by depositing thin film of plasmonic materials on the patterned structure created by template/mask. The second method is by depositing a thin film of plasmonic materials directly on the nanostructured substrate. It is common to design a nanostructured substrate using template/mask and etching the substrate to generate the pattern. The third method involves depositing a thin film of plasmonic materials on a planar surface followed by adding/subtracting features/height of the deposited thin film. Finally, the fourth method involves transferring the already plasmonic active films or individual nanostructures on the substrate. Unsupported PANTFs are of less interest for biological applications because of physical instability; therefore, they are commonly prepared or transferred on the surface of inert substrates, such as glass, silicon wafers, and mica for applications.

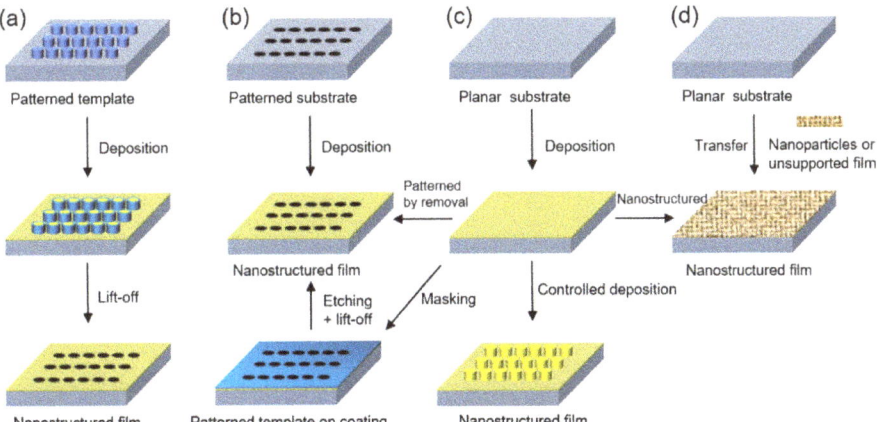

Figure 2. Schematic of commonly applied strategies for the synthesis of plasmonic-active nanostructured thin films starting from (**a**) deposition on patterned template, (**b**) deposition on patterned substrate, (**c**) deposition on planar substrate, and (**d**) transfer on planar substrate.

5.1. Patterned Template for PANTFs Synthesis

Varieties of templates have been used on the substrate along with different deposition techniques to design the desired PANTFs. The general approach for the fabrication of many template-based PANTFs are nearly the same, which involve creating a patterned structure on top of the substrate and deposition of the materials of interest on or around the patterned structures. Some methods require removing the template or mask after deposition to create the PANTFs, whereas others create the PANTFs without the removal of the template or mask. Here we discuss some of the most used template-based PANTF synthesis methods.

5.1.1. Electron Beam Lithography (EBL)

It is one of the most powerful techniques for nanostructure formation with nanometer-scale precision in designing shape, size, and arrangement [37,38]. Consequently, the surface plasmon resonance peak can be tuned over wide wavelengths. The typical steps in the EBL technique involve spin-coating of a thin film of resist on a solid support, designing desired patterns on resist using a beam of electrons, developing the resist, depositing the desired metal, and finally, removing the template (lift-off) [39]. The exposure of electrons beams to resist change its solubility allowing selective removal of either the exposed or non-exposed region. The commonly used positive resist for EBL is poly(methyl methacrylate) (PMMA) which can be developed (solubilize) using a mixture of methyl isobutyl ketone (MIBK) and isopropyl alcohol and removed during lift-off step using acetone. EBL has been widely used for creating a range of plasmonic-active arbitrary nanostructures as discrete arrays or films. The nanostructures designed by EBL have better reproducibility compared to random nanostructures prepared using different deposition techniques. They are also suitable for modeling experiments and understanding properties of nanostructure with the controlled variation of shape, size, and spacing.

Despite its many advantages, the EBL technique for PANTF synthesis is time-consuming, costly, and requires expertise in the field [8]. Moreover, the substrate must be conductive to avoid charging effect. A thin layer of conductive material (e.g., ITO, Au, or Al) is commonly deposited in between the substrate and resist or on top of resist to achieve conductivity. The nanostructure formed is less sensitive due to plasmonics damping than the nanostructures prepared using other methods. These limitations of EBL hindered the large-scale fabrication of PANTFs.

5.1.2. Nanosphere Lithography

Nanosphere lithography (NSL) was developed to overcome the shortcoming of EBL techniques. It is a simple, cost-effective, and high-throughput periodic nanostructure fabrication technique capable of designing a wide variety of nanostructures [40]. In a typical NSL fabrication method, a suspension of polystyrene or silica nanospheres is drop or spin-cast on the substrate to form a hexagonally close-packed self-assembled monolayer [41]. The distance between the nanospheres can be controlled by different means. One method is to etch the hexagonally close-packed nanosphere using reactive ion etching (RIE) until the desired gap between the nanosphere is obtained [42]. The other method includes electrostatically separating nanospheres by controlling the concentration of the salt in the colloidal solution [43]. The former method can still create an array of periodically patterned nanostructures, whereas the latter may not have perfect periodicity. The monolayer of the nanospheres can now act as a substrate for the direct deposition of film over spheres creating PANTFs. It also acts as a mask for the deposition of plasmonic materials. After the deposition, the polystyrene spheres can be easily removed using physical methods (e.g., tape stripping, sonicating, etc.) or treating with organic solvents (e.g., absolute ethanol, chloroform, DCM, toluene, etc.) to create PANTFs. In the NSL technique, the peak wavelength of LSPR can be tuned by changing the diameter of nanospheres, changing the distance between nanospheres and by varying deposition parameters such as time, angle, etc. However, the NSL strategy is not free from shortcomings. The monolayer of nanospheres can easily create configurational disorder leading to poorly reproducible PANTFs. Thermal evaporation of plasmonic materials should be avoided for deposition, as the increase in temperature can easily damage polystyrene and its arrangement.

5.1.3. Nanoimprint Lithography

The nanostructured mold can be prepared once using techniques having better resolution but could be expensive and time-consuming including EBL, NSL, etc. As-prepared mold can then be pressed and attach to the resist on substrate transferring the nanostructured features of the mold to the resist, called nanoimprint lithography. The curing of the resist is done using UV or heating, followed by removal of the mold [44]. After etching, the desired plasmonic materials can be deposited followed by lift-off of resist to obtain desired PANTFs. The same mold can then be used multiple times for creating copies of exact same PANTFs saving time, cost, and efforts.

5.1.4. Porous Membrane-Based Lithography

Different types of nanoporous thin structures can be fixed on top of a substrate and can be used as a template for depositing plasmonic materials creating PANTFs. Thin-film porous aluminum oxide is one of the commonly used templates for creating PANTFs [45,46]. A typical approach involves depositing a thin film of plasmonic materials on the substrate followed by aluminum. The anodization of the aluminum to anodic aluminum oxide (AAO) creates high-density arrays of nanopores. The diameter and depth of the nanopores can be tuned by varying the anodizing potential and aluminum deposition time, respectively. The PANTFs can be created by directly depositing plasmonic materials on thin film of plasmonic materials through holes of AAO by varying different parameters or by removing the AAO template after deposition to create nanopillars [47,48].

5.2. Patterned Substrate for PANTFs Synthesis

Patterns on the substrate can be prepared by masking the portion of the substrate and etching from the exposed area. As-prepared patterned substrates are suitable for nanoimprint lithography. However, the plasmonic materials can also be directly deposited to the patterned surface to obtain PANTFs.

5.3. Planar Thin Films for PANTFs Synthesis

The PANTFs can also be prepared by starting from a deposited thin film of plasmonic materials. Three different approaches to turn the thin film into PANTFs are by removing the deposited materials, adding materials, and annealing thin films into nanostructures with some kind of force without removing any materials.

5.3.1. PANTF by Removal of Material

The simplest approach to create PANTFs from thin films with the removal of thin film material but without the use of resist or mask is by using focused ion beam (FIB) milling. The FIB can be used to construct the desired shape or pattern by directly focusing the ion-beam on thin films. The advantage of this method is that it does not require any template to create nanostructure as in EBL and NSL. This reduces the tedious fabrication steps. However, this is still time-consuming and expensive for large scale production. Additionally, the heavier ions, most commonly gallium ions, used for creating the nanostructures easily contaminate the sample by implanting the heavy metal and have low resolution compared to EBL [37]. However, with the recent development of technologies, helium ion beams can now be used to create a nanostructured thin film with a resolution of 3.5 nm [49].

Depositing thin metals alloy film and dissolving the less noble metals from the alloy using chemical or electrochemical techniques can create PANTFs. Alternatively, thin metal layers can be deposited, followed by annealing to mix and dealloying to create porous nanostructures. Nanoporous gold and silver thin films can be created using this method.

5.3.2. PANTF by Addition of Material

In general, depositing materials by changing the parameters of vapor and sputtering techniques does not create suitable nanostructures on top of thin films without the use of templates. However, wet chemical deposition techniques, especially the electrochemical method, can produce highly sensitive nanostructures on the thin films with or without the use of structure-directing agents [50]. The variation in potential, current, and time can produce different structures. Additionally, this method does not require high temperature, pressure, and vacuum as needed for EBL and FIB techniques. Although the PANTFs created by electrochemical methods are highly sensitive, they are randomly oriented in the absence of a template making it difficult for systematic study.

5.4. Transfer of Film and Nanoparticles

Plasmonic-active nanostructures can be prepared in a solution phase using various strategies. The common approach is to reduce the metal salts into nanostructures in the presence of stabilizing agents. Various nanostructure directing agents and additives can be used to prepare the desired shape and size. Similarly, plasmonic-active unsupported thin-films can also be prepared on solution either using a bottom-up or top-down approach. These nanostructures and thin films can be easily transferred to the substrate to create supported PANTFs. Diverse types of block copolymers can be used to arrange nanoparticles in the desired orientation.

6. Characterization of PANTFs

6.1. Sensitivity

The sensitivity of PANTFs in term of SPP and LSPR is commonly reported as bulk refractive index sensitivity (RIS), which is an output response per refractive index unit. For LSPR, the change in peak wavelength of extinction spectra (λ_{max}) is directly proportional to change in refractive index (n) and is given by Equation (1) [4].

$$\Delta\lambda_{max} = m\Delta n[1 - exp(\frac{-2d}{l_d})] \qquad (1)$$

where m is the bulk refractive index response, d is the thickness of the adsorbate layer, and l_d is electromagnetic field decay length. However, the sensitivity of PANTFs should be more accurately represented as a figure of merit (FOM), which is bulk RIS per full width at half maxima (FWHM), Equation (2).

$$\text{FOM} = \frac{\text{Bulk RIS}}{\text{Full width at half maxima (FWHM)}} \qquad (2)$$

When comparing the sensitivity of different PANTFs, care should be given to different structural and composition parameters. In general, a PANTF showing initial LSPR peak wavelength at NIR has higher refractive index sensitivity than the PANTF that has initial LSPR peak wavelength at the visible region. Therefore, these two structures should not be compared to prove better PANTFs. The FOM should be included along with the bulk refractive index sensitivity for the comparison of PANTFs to prove structural improvement. Unlike SPP and LSPR, SERS sensitivity is reported based on enhancement factor (E), which is highest when the plasmon wavelength is between the Raman excitation and emission energies [4,51]. Table 1 presents the different types of plasmonic-active PANTFs with the fabrication methods and their sensitivity.

Table 1. Sensitivity of plasmonic-active nanostructured thin films.

NS	M	Fabrication Method	d/p/t (nm)	λ_{max} (nm) Air or N_2	Method	Sensitivity R = (nm/RIU), FOM, E	Ref
	Al	EBE/EBL/Etching	220/-/100	510	LSPR	R = 487	[52]
		NH Si template/EBE/peel off	180/500/100	NA	LSPR	R = 450	[44]
		NH Si template/EBE/peel off	100/500/200	NA	LSPR	R = 494	[53]
		NSL/RIE/deposition	290/-/80	550/790	LSPR	R = 252	[54]
	Ag	NSL/RIE/deposition/HF etching	290/-/80	655	LSPR	R = 648	
		deposition/FIB	200/545/100	645	LSPR	R = 400	[55]
		NSL/RIE/EBE/lift-off	~300/400/50	672	SERS	E = 8.13 × 10^5	[56]
		NSL/RIE/EBE/lift-off/plating	<300/400/>50	NA	SERS	E = 3 × 10^6	
NH		NSL/RIE/EBE/lift-off	60/-/20	~620	LSPR	R = ~70	[57]
		NSL/RIE/EBE/lift-off	60/-/20	675 ± 10 (1 hole)	LSPR	R = ~90	
		NSL/PVD/lift-off	60/-/20	575	LSPR	R = ~100	[43]
		UV-NL/RIE/EBE/lift-off	200/400/50	583	LSPR	R = 150	[58]
	Au	NSL/RIE/EBE/lift-off	70/	NA	SPP	R = >3000	[59]
		NSL/RIE/EBE/lift-off	500/-/80	1489	LSPR	R = 375	[60]
		NSL/EBE/mask/RIE/mask removal/lift-off	500/-/80	898	LSPR	R = 625	
		NSL/RIE/sputtering/lift-off	600/1000/125	710	LSPR	R = 530 ± 30 FOM = 132	[11]
			600/1000/125	NA	SPP	R = 3600 ± 200 FOM = 327	
		EBL/PVD	100/585/50	NA	SERS	E = ~10^6	[61]
		NSL/PVD/lift-off	125/-/40	575	LSPR	R = 36	[62]

Table 1. Cont.

NS	M	Fabrication Method	d/p/t (nm)	λ_{max} (nm) Air or N_2	Method	Sensitivity R = (nm/RIU), FOM, E	Ref
NPi	Au	porous Al_2O_3 imprinted nanopillars of Cyclo-olefin polymer/sputter	30.0–39.9/-/50	NA	LSPR	R = 154	[63]
		Au thin film/Porous Al_2O_3/ED	Film t = 5 nm 25/60/380	NA	SPP	R = 32,000 FOM = >330	[48]
	Al	LIL/deposition	180/400/150	310	SPP	R = 223 FOM = 8	[64]
			180/400/150	413	SPP	R = 485 FOM = 20	
NC	Al	EA (120 V)/removal Al_2O_3	/246.3/	250	LSPR	R = 191	[65]
		EA (195 V)/removal of Al_2O_3	/456.7/	350	LSPR	R = 291	
FON	Al	NSL/PVD	210/-/200	NA	SERS	E = ~10^4–10^5	[66]
NPA	Cu	NSL/EBE	100/-/30	580	LSPR	R = 67.8	[67]
HA	Au	THL/EBE	620/620/100	NA	SPP	FOM = 730	[68]
NGF	Au	Au thin film/ED	100–200/-/200	518 ± 1	LSPR	R = 100 ± 2	[50]
NPG	Au	Dealloyed unsupported thin film in HNO_3/transferred to substrate	30/-/100	~510	LSPR	R = 210	[69]
			50/-/100	NA	LSPR	R = 264	

Notes: NS = Nanostructures; NH = nanohole; NPi = nanopillar; NC = nanoconcave; FON = film over nanosphere; NPA = nanoparticle array; HA = hexagonal array; NGF = nanostructured gold film; NPG = nanoporous gold; RIS = bulk refractive index sensitivity; E = enhancement factor; EBE = electron beam evaporation; PVD = physical vapor deposition; EBL = electron beam lithography; LIL = laser interference lithography; THL = tunable holographic lithography; ED = Electrochemical deposition; NL = nanoimprint lithography; EA = electrochemical anodization; d = diameter of hole; p = periodicity and t = thickness of the film; FOM = figure of merit.

6.2. Imaging

The morphologies of PANTFs can be characterized using scanning electron microscopy (SEM), which helps to determine the diameter and periodicity of nanostructures on a thin film. The thickness of the PANTFs can be determined using atomic force microscopy (AFM). Dark-field optical microscopy (DFM) is frequently used for analyzing larger nanostructures (above 100 nm) on thin film [43]. The DFM is also handy to study single nanostructure on thin film [57].

6.3. Simulations

Finite-difference time-domain (FDTD) calculations were commonly performed on PANTFs to obtained theoretical plasmonic spectra and are compared with experimental spectra [56]. 3D FDTD simulation of the time-averaged surface plasmon field intensity can be generated around the edges of nanostructures. This helps to monitor the change in size and position of the hot spots with the change in different parameters such as with or without a coating of a thin silica layer on PANTFs [44].

7. Types of PANTFs

A wide variety of nanostructured thin films have been synthesized based on the methods discussed earlier. However, the general approach of synthetic method is either top-down or bottom-up. In top down approach, the PANTFs are created after removal of some part of the original material, whereas in the bottom-up approach material is added. Similar structures can be created using both approaches but with different strategies. In this part, we discuss the strategies for fabricating some of the most common types of PANTFs. More specifically, we will discuss PANTFs having three uniquely distinguishable features, (1) thin films with gaps throughout the film (e.g., nanoholes) and (2) thin films with the elevated nanofeatures throughout the films (e.g., nanopillars, nanodomes, nanospikes, etc.), and (3) transferred films

7.1. PANTFs with Gaps

One of the fundamental constraints for manipulating light is that its transmittivity through the apertures smaller than the wavelength of the photon is extremely low. In 1998, Ebbesen et al. were able to transmit photons having wavelengths as large as ten times the diameter of the hole by designing the array of sub-wavelength circular holes on 200 nm thick silver film [70]. This was due to the coupling of light with surface plasmons on plasmonic-active nanohole array thin film. They used the FIB technique to create an array of holes. Zhu and Zhou used FIB to create pentagram nanohole arrays, which show 2.4 times improvement in the maximum transmission compared to the traditional transmission due to the excitation of surface plasmons. Recently, many other techniques have been developed and improved to create nanohole arrays and are used for various applications. Nevertheless, FIB is still improving and is widely used for preparing nanoholes in varieties of the shape because of its nanometer resolution and template-free nanostructure forming capacity [49].

The transmission peak wavelength and refractive index sensitivity of the nanohole array depend on the diameter and periodicity of the hole as well as the thickness and composition of the film [71]. With the increase in diameter of the holes or distance between the holes, the plasmon resonance peak red shifts [43]. The LSPR bulk refractive index sensitivity of Au nanohole array has been reported from as low as 36 nm/RIU [62] to hundreds of nanometers [60].

NSL is one of the most used methods for creating nanohole array by controlling different parameters. Figure 3a shows the strategy to create two different types of silver nanohole arrays using NSL [56]. This technique involves three important steps, first is reactive ion etching of surface arranged polystyrene, second is the deposition of the substrate with or without tilting a holder, and final is the removal of the polystyrene. The shape of the nanohole was found to vary with change in the angle during metal deposition and hence the sensitivity of the nanostructures. Figure 3b shows SEM images of hexagonal nanohole array prepared at 0° tilt angle (top) and elliptical nanohole array prepared at 60° tilt angle (middle). Figure 3b bottom is a nanohole array after electroless plating with Ag. Figure 3c(A) is a SERS spectrum of nanohole array plated for 4 min with Ag. It shows that as-plated nanohole array gives better SERS enhancement (enhancement factor = 3×10^6) than 4 min Ag plated flat silver film (B), a standard nanohole array (C), and a flat silver film (D).

A single nanohole in the array can show better LSPR sensitivity than a dense hole sample. A work by Rindzevicius et al. found that Au nanohole (diameter = 60 nm) array prepared using NSL show LSPR sensitivity of ~70 nm/RIU, which is low compared to the sensitivity of a single hole of the same array ~90 nm/RIU [57]. They used the single nanohole for the study of successive molecular adsorption processes. They also showed that the dense hole array can be used for long-range refractive index sensing up to 100–340 nm preparing Langmuir–Blodgett multilayers composed of 22-tricosenoic acid [12]. NSL has also been used to prepare 3D-nanohole array, which shows better LSPR response (1.7-fold enhancement) compared to regular 2D-nanohoe array [60].

Owing to the plasmonic sensitivity of nanoholes, they have been used for studying different biomolecules and their interaction. The unique advantage of the nanohole array to other nanostructures is that it can be designed to penetrate through the substrate creating nanofluidic channels [72]. A biosensor chip having nanohole of 150 nm penetrating through 250 nm gold and silicon nitride layer supported on Si wafer was designed using NSL to study the interaction between biotin and NeutrAvidin [72]. Nanohole arrays prepared using the NSL technique show high SPP sensitivity of >3000 nm/RIU (2-fold improvement compared to thin films) [59]. It was used for detecting immunoglobulin G with the detection limit 10 nM.

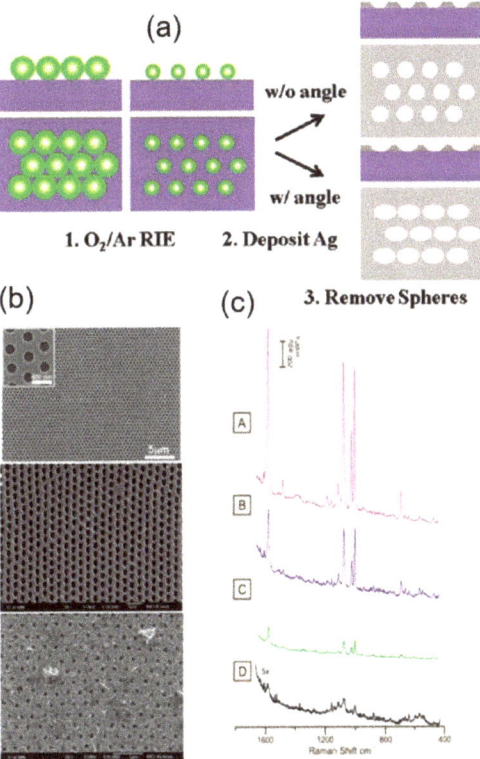

Figure 3. (**a**) Schematic representation of process steps for fabricating nanohole arrays using nanosphere lithography. (**b**) SEM images of (top) the resulting large area (~30 × 30 µm^2) single crystalline hexagonal nanohole array, (middle) an elliptical nanohole array (60° tilt angle) after nanosphere removal, and (bottom) after 4 min Ag electroless plating onto the nanohole array. (**c**) SERS spectra of (**A**) 4 min Ag plated nanohole array, (**B**) 4 min Ag plated flat silver film, (**C**) a standard nanohole array, and (**D**) 5× a flat silver film. All substrates were exposed to 1 mM benzenethiol prior to SERS measurements. P = 0.32 mW, t = 180 s, and λ_{max} = 532 nm. Reproduced (**a**,**c**) and adapted (**b**) with permission from Reference [56], Copyright 2009, American Chemical Society.

The EBL is another popular technique for making nanohole arrays applying different strategies. One of the strategies involves direct deposition of a thin film of plasmonic material on top of a patterned nanohole obtained after the development of resist. The PMMA works as a support for the nanohole array and is not removed. Recently, this strategy was utilized by Luo et al. for preparing plasmonic-active nanohole array having SERS enhancement factor (E) = ~10^6 and used it to determine DNA methylation [61]. EBL can also be used for etching and creating nanohole array on substrate, which can then be used for creating plasmonic active thin film nanohole array by direct deposition of plasmonic material. EBL is frequently used to create arrays of nanopillar that can then be used as a nanoimprint stamp for creating ordered nanohole array on resists.

A master template of nanohole array can be fabricated on the substrate using different lithography techniques. This master template can be used for depositing a thin film of plasmonic material and template-stripped on to another substrate [73]. After stripping, the master template can be repeatedly used for making many sets of nanohole array. Therefore, this method avoids needing to repetitively prepare the template from the beginning using lithographic techniques, saving time and cost. Figure 4A shows schematic of preparing nanohole array based on the template-stripping method. The steps include designing a nanohole array on resist film using nanoimprint stamp and etching from the holes to obtain Si master template of nanohole array. The template can then be coated with a metal film, which can be stripped from the template to create template-stripped plasmonic-active nanohole arrays thin film. The SEM image of Si template Figure 4B(a), after Ag deposition Figure 4B(b), and template stripped plasmonic-active nanohole arrays thin film Figure 4B(c) were shown along with the photographic images of fabricated nanohole array chip Figure 4B(d) and multi-channel PDMS chip attached on the silica-coated nanohole array chip (inset in Figure 4B(d)).

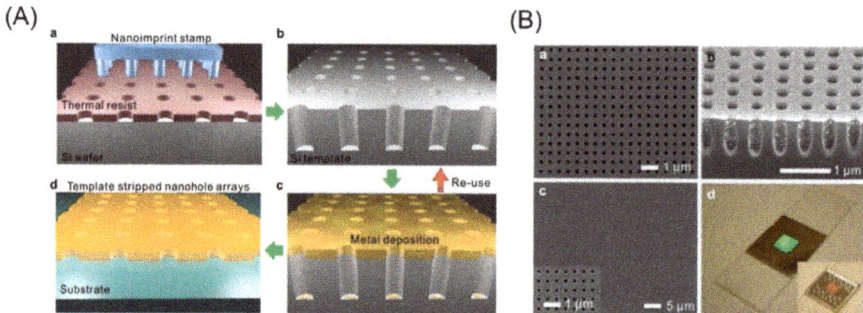

Figure 4. (**A**) Schematic for fabricating the large-area nanohole arrays. (**a**) A thermal resist layer spun on a Si wafer is imprinted with a nanoimprint stamp with circular post patterns. (**b**) The Si wafer is subsequently etched to be a nanohole template with deep circular trenches. (**c**) A metal film is directionally deposited on the Si template. (**d**) The metal surface is coated with a thin layer of epoxy and covered with a glass slide. The Ag film is then peeled off of the template to reveal the smooth nanohole array made in the metal film. The Si template can be reused to make multiple identical samples. (**B**) (**a**) A SEM image of the Si template with deep circular trenches. (**b**) A cross-sectional SEM image of the Si template after depositing a 100 nm thick Ag film. (**c**) A SEM image of the template-stripped Ag periodic nanohole array. The inset shows a zoomed-in image of the template-stripped Ag nanoholes. The diameter of the nanoholes and periodicity of the array are 180 and 500 nm, respectively. (**d**) A photograph of the fabricated nanohole array chip. A 26.5 mm × 26.5 mm area of 100 nm-thick Ag film with nanohole patterns in an 8 mm × 8 mm area in the center is transferred to a standard microscope slide. The inset in panel d shows a photograph of a multi-channel PDMS chip attached on the silica-coated nanohole array chip. Reproduced with permission from Reference [44], Copyright 2011, American Chemical Society.

7.2. PANTFs with Elevated Nanofeatures

The PANTFs with the elevated nanofeatures can be prepared using different fabrication strategies. Broadly, fabrication strategies can be grouped into template-based and template-free techniques. Reflectance spectroscopy is frequently used for measuring the response from these elevated PANTFs when the light cannot transmit throw the film.

7.2.1. Film over Nanospheres

One of the simplest approaches to create elevated PANTFs is by directly depositing plasmonic materials over hexagonally close-packed nanospheres to create a film over nanospheres (FON). By tuning the size of nanospheres and thickness of the deposited plasmonic materials, the LSPR spectra of FON can be tuned across the entire visible region [74]. The PANTFs prepared using this method can also highly enhance the SERS signals. Zhang et al. prepared very stable AgFON substrates to quickly detect anthrax spores [75]. Calcium dipicolinate, a biomarker for bacillus spores, was extracted and detected using SERS within 11 min with a limit of detection of ~2.6×10^3 spores. The same group later modified the AgFON with atomic layer deposition of alumina layer improving the stability of the AgFON from nearly 1 month to 9 months and improving the detection limit of anthrax spores to ~1.4 × 10³ spores [76]. Masson et al. etched the hexagonally close-packed nanosphere before the deposition to tune the gap between the spheres [77]. They found that optimal SERS response can be obtained when a ratio of a gap to the diameter of a sphere is less than 1. They also observed that the optimal excitation wavelength and roughness improve SERS response. They also studied the role of the position of a metal layer on bimetallic films. SERS response was improved when Au layer was prepared on top of Ag layer but decreased when Ag layer was prepared on top of Au layer. Improved structure to the bimetallic films could be preparing a hybrid structure of AgFON and Au nanoparticle as shown in Figure 5 [78]. Figure 5a–c are SEM images of (a) AgFON prepared on 505 nm polystyrene bead, (b) AuNP-AgFON hybrid structure, and (c) cross-section of (b). Figure 5d is SERS spectra of benzenethiol from AuNP-AgFON-505, AgFON-505, and AuNP-Ag film showing three prominent Raman bands. The AuNP-AgFON-505 shows ~30 times larger SERS enhancement compared to AgFON-505, indicating the significant improvement of the SERS response. Figure 5e–f are Raman mapping images of AuNP-AgFON-505 and AgFON-505 at 785 nm excitation laser, respectively. Figure 5g–g2 are FDTD simulations of E-fields on AuNP assemblies on AgFON-505, zoomed in image for dimer AuNPs, and crevice gap, respectively, in AuNP-AgFON-505 system.

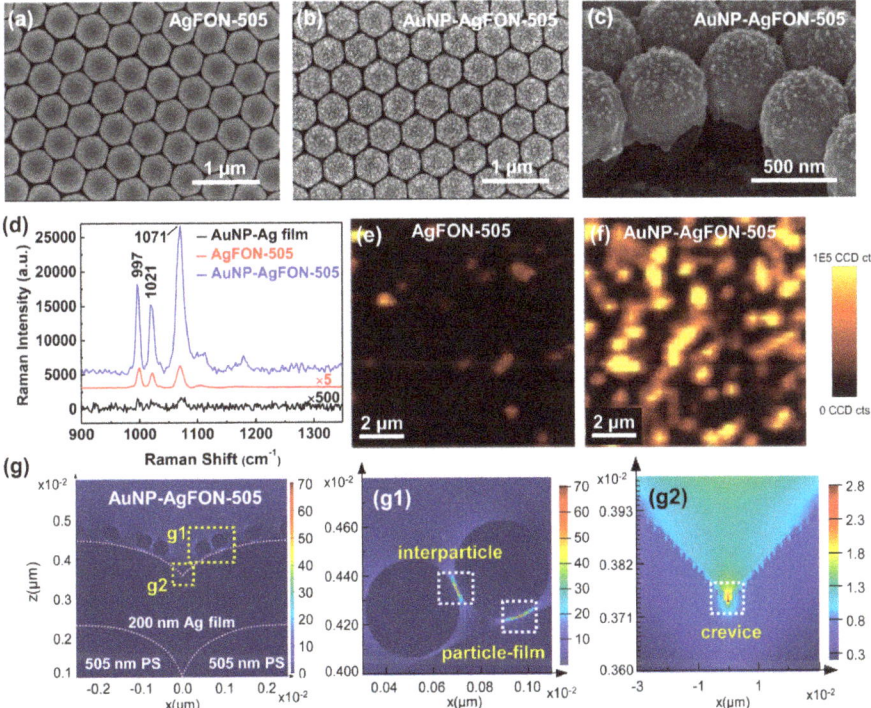

Figure 5. SEM images of AgFON-505 (**a**), AuNP-AgFON-505 (**b**), with cross-sectional image of AuNP-AgFON-505 (**c**). Raman spectra of AuNP-AgFON-505, AgFON-505, and AuNP-Ag film substrates (**d**). Raman mapping images of AuNP-AgFON-505 versus AgFON-505 at 785 nm excitation laser (**e**,**f**). (**g**) FDTD simulations of E-fields on AuNP assemblies on AgFON-505 substrate at 785 nm laser wavelength. Partially enlarged image for dimer AuNPs (**g1**) and crevice gap (**g2**) in AuNP-AgFON-505 system. Reproduced with permission from Reference [78], Copyright 2016, American Chemical Society.

7.2.2. Array of Nanodomes and Nanopillars

One way to prepare these types of PANTFs is by one step deposition of plasmonic materials on nanofeatured substrate. However, even without making nanofeatures on the substrate, these types of structures can be obtained using a template and two-step deposition. This technique involves the deposition of a thin planar film of plasmonic material on substrate as a first deposition. The second deposition is carried out after patterning the film with resists or nanospheres. Finally, the elevated array is achieved after lifting off the resist and nanospheres. Figure 6a is a schematic diagram of the NSL technique to design nanopatterns on the substrate followed by one step deposition to obtain PANTF. Figure 6b(A,B) are low and high magnification SEM images of the Ag deposited film, respectively. Figure 6c(A,B) show changes in LSPR peak wavelength with the change in bulk refractive index around PANTFs.

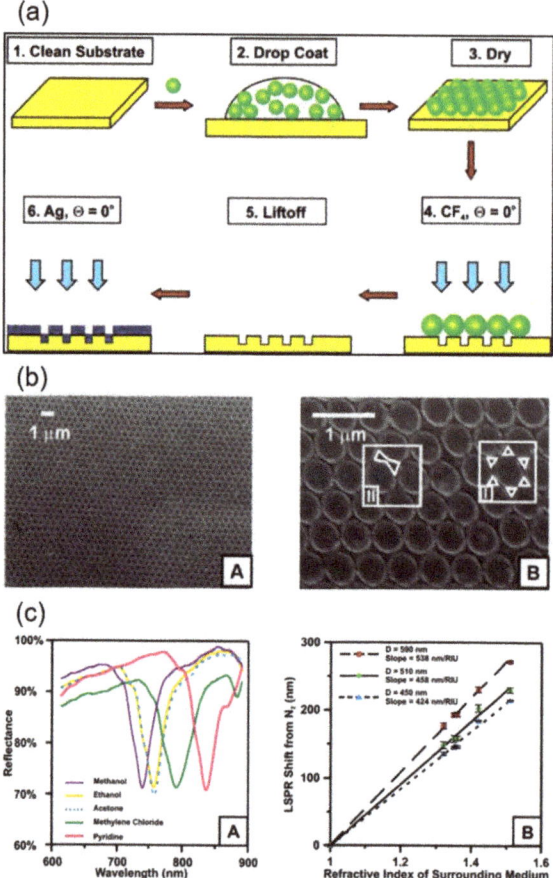

Figure 6. (a) Schematic illustration of preparation of the film over nanowell surfaces. (b) SEM images of Ag film over nanowell surface [diameter of nanospheres (D) = 510 nm; mass thickness of Ag film (d_m) = 50 nm; and etch time (t_e) = 10 min]. (A) Well-packed area of over 40 μm^2 and (B) magnified image of the same sample. The SEM accelerating voltage was 5 kV. (c) (A) Collection of reflectance spectra of Ag film over nanowell surface in different solvents (D = 590 nm; d_m = 50 nm). (B) Plots of λ_{min} (solvent)–λ_{min} (dry nitrogen) versus refractive index of the solvent for three nanosphere sizes: D = 450, 510, and 590 nm. Each data point represents the average value obtained from at least three surfaces. Error bars show the standard deviations. For all surface preparations, d_m = 50 nm and t_e = 10 min. Reproduced with permission from Reference [41], Copyright 2005, American Chemical Society.

7.2.3. Randomly Oriented Nanospikes and Nanobricks

It is well known that random nanostructures have better LSPR sensitivity and SERS enhancement compared to their ordered counterparts. Electrochemical methods can create randomly oriented PANTFs on a conductive substrate without the use of a template. A review of electrochemical methods for preparing thin nanoporous gold films has been covered in an article previously [79]. The PANTFs can be prepared by either by adding different structure-directing agent or varying the deposition parameters. SEM images in Figure 7a,b show two completely different morphology of Au nanostructures prepared as a film with or without adding structure-directing agent. Figure 7a shows SEM and AFM images of a gold nanospike thin film prepared by providing a constant potential of 0.05 V versus Ag/AgCl (3M KCl) at different deposition times from HAuCl$_4$ solution containing Pb (CH$_3$COO)$_2$ as a structure directing agent [80]. With the increase in deposition time from 360 s to 540 s, spike height increases drastically to 302 ± 57 nm. The shapes and size of nanostructures can also be tuned by changing the concentration of structure-directing agent. Figure 7b(A) is an SEM image of nanostructured gold film (NGF) prepared by providing −1.2 V for 60 s followed by −1.6 V for 30 s from 50 mM potassium dicyanoaurate solution [50]. Figure 7b(B) shows a typical reflection-based LSPR biosensing setup, where the fiber optic probe is used for both directing incident light for the excitation of surface plasmons and collecting the reflected light. The bulk refractive index sensitivity of the as-prepared NGF was determined by changing the environment of NGF with different concentrations of glycerol as shown in Figure 7b(C) and was found to be 100 ± 2 nm/RIU with FOM of 1.7. Figure 7b(D) shows real-time interactions of surface-immobilized carbohydrate mannose with different concentrations of lectin concanavalin A on NGF surface using reflection-based LSPR spectroscopy.

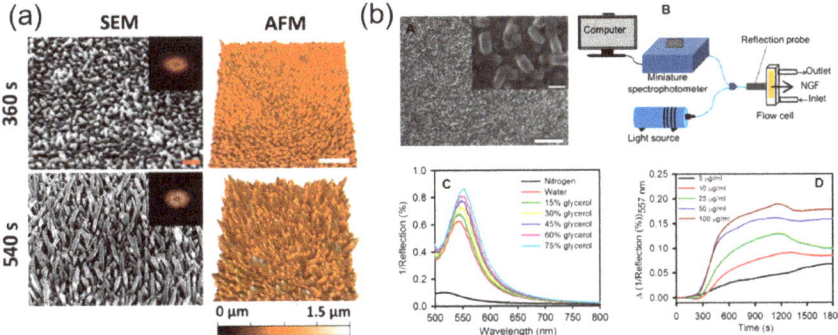

Figure 7. (**a**) Native surface data for the nanospike deposition from precursor solution of 6.8 mM HAuCl$_4$ and 1 mM Pb(CH$_3$COO)$_2$: column (1) SEM (45° tilted) and column (2) AFM images of the gold nanospike surfaces as a function of electrodeposition time. The red and white scale bars are 200 nm and 1 mm, respectively. 2D-FFT data (insets) are shown for the top-down SEM images. Adapted with permission from Reference [80], Copyright 2018, The Royal Society of Chemistry. (**b**) (**A**) SEM images of nanostructured gold film (NGF) prepared by providing −1.2 V for 60 s followed by −1.6 V for 30 s (vs. Ag/AgCl (KCl, Satd) from 50 mM potassium dicyanoaurate. Scale bars: 2 µm. Insets are the corresponding higher magnification SEM images (scale bars: 0.2 µm). (**B**) Optical set up for localized surface plasmon resonance spectroscopy in reflection mode. (**C**) Bulk refractive index response of as-prepared NGF evident by change in LSPR peak wavelength. (**D**) Real-time LSPR response of self-assembled monolayer (SAM)-modified NGF to different concentrations of Concanavalin A. NGF was surface modified with a mixed SAM of αMan-C8-SH and TEG-SH (1:3). Reproduced with permission from Reference [50], Copyright 2014, Elsevier Ltd.

7.3. Transferred Films

The plasmonic-active individual nanoparticles and unsupported plasmonic-active thin films can be directly transferred to the solid supports creating PANTFs. Lang et al. prepared 100 nm thin film of nanoporous gold (np-Au) by chemically dealloying gold alloy film in 70% HNO_3 [69]. The size of the pores of np-Au was controlled by varying the dealloying time from 5 min to 24 to create 10 to 50 nm. As-prepared np-Au thin film was transferred to glass slides to obtain the PANTFs with bulk refractive index sensitivity of nearly 210 nm/RIU and 264 nm/RIU for 30 and 50 nm pores, respectively.

8. Conclusions

The PANTFs have garnered huge attention in the field of plasmonics because of their ability to couple light with surface plasmons causing simultaneous excitation of SPP and LSPR. This can have a huge implication for improving and miniaturization of plasmonic and photonics-based devices as well as integrating them with microfluidics, quartz crystal microbalance, and electrochemical devices. In this review, we discussed propagating and localized surface plasmon resonance and recent advances in the synthetic methodologies for PANTFs. We also reviewed the advantages of using PANTFs compared to discrete nanostructures along with the comparison of the sensitivity of different types of PANTFs.

The discovery of PANTFs has led to the integration of plasmonics with quartz crystal microbalance, microfluidics, and electrochemistry. However, more work is needed in the field for improving sensitivity and simplifying the system for commercial use. As the research on plasmonics is expanding quickly, we can expect to see the integration of all the four fields in a single system in the near future.

Author Contributions: J.K.B.—conceptualization and original draft preparation; M.H.U.M.—review and additional writing; K.J.S.—Review, additional writing, editing, and funding acquisition. All authors have read and agreed to the published version of the manuscript.

Funding: The authors acknowledge the recent support of the work in this area by the University of Missouri–St. Louis and by the NIGMS awards R01-GM111835.

Conflicts of Interest: The authors declare no conflict of interest.

References

1. Jauffred, L.; Samadi, A.; Klingberg, H.; Bendix, P.M.; Oddershede, L.B. Plasmonic heating of nanostructures. *Chem. Rev.* **2019**, *119*, 8087–8130. [CrossRef] [PubMed]
2. Tokel, O.; Inci, F.; Demirci, U. Advances in plasmonic technologies for point of care applications. *Chem. Rev.* **2014**, *114*, 5728–5752. [CrossRef] [PubMed]
3. Zeng, S.; Baillargeat, D.; Ho, H.-P.; Yong, K.-T. Nanomaterials enhanced surface plasmon resonance for biological and chemical sensing applications. *Chem. Soc. Rev.* **2014**, *43*, 3426–3452. [CrossRef]
4. Willets, K.A.; Van Duyne, R.P. Localized surface plasmon resonance spectroscopy and sensing. *Annu. Rev. Phys. Chem.* **2007**, *58*, 267–297. [CrossRef]
5. Homola, J. Surface plasmon resonance sensors for detection of chemical and biological species. *Chem. Rev.* **2008**, *108*, 462–493. [CrossRef]
6. Mayer, K.M.; Hafner, J.H. Localized surface plasmon resonance sensors. *Chem. Rev.* **2011**, *111*, 3828–3857. [CrossRef]
7. Jiang, N.; Zhuo, X.; Wang, J. Active plasmonics: Principles, structures, and applications. *Chem. Rev.* **2017**, *118*, 3054–3099. [CrossRef]
8. Kang, H.; Buchman, J.T.; Rodriguez, R.S.; Ring, H.L.; He, J.; Bantz, K.C.; Haynes, C.L. Stabilization of silver and gold nanoparticles: Preservation and improvement of plasmonic functionalities. *Chem. Rev.* **2018**, *119*, 664–699. [CrossRef]
9. Shao, L.; Susha, A.S.; Cheung, L.S.; Sau, T.K.; Rogach, A.L.; Wang, J. Plasmonic properties of single multispiked gold nanostars: Correlating modeling with experiments. *Langmuir* **2012**, *28*, 8979–8984. [CrossRef] [PubMed]
10. Barbosa, S.; Agrawal, A.; Rodriguez-Lorenzo, L.; Pastoriza-Santos, I.; Alvarez-Puebla, R.A.; Kornowski, A.; Weller, H.; Liz-Marzan, L.M. Tuning size and sensing properties in colloidal gold nanostars. *Langmuir* **2010**, *26*, 14943–14950. [CrossRef] [PubMed]

11. Couture, M.; Live, L.S.; Dhawan, A.; Masson, J.F. EOT or Kretschmann configuration? Comparative study of the plasmonic modes in gold nanohole arrays. *Analyst* **2012**, *137*, 4162–4170. [CrossRef] [PubMed]
12. Rindzevicius, T.; Alaverdyan, Y.; Kaell, M.; Murray, W.A.; Barnes, W.L. Long-range refractive index sensing using plasmonic nanostructures. *J. Phys. Chem. C* **2007**, *111*, 11806–11810. [CrossRef]
13. Bryche, J.F.; Gillibert, R.; Barbillon, G.; Gogol, P.; Moreau, J.; de la Chapelle, M.L.; Bartenlian, B.; Canva, M. Plasmonic enhancement by a continuous gold underlayer: Application to SERS sensing. *Plasmonics* **2016**, *11*, 601–608. [CrossRef]
14. Henzie, J.; Lee, J.; Lee, M.H.; Hasan, W.; Odom, T.W. Nanofabrication of plasmonic structures. *Annu. Rev. Phys. Chem.* **2009**, *60*, 147–165. [CrossRef] [PubMed]
15. Willets, K.A.; Wilson, A.J.; Sundaresan, V.; Joshi, P.B. Super-resolution imaging and plasmonics. *Chem. Rev.* **2017**, *117*, 7538–7582. [CrossRef] [PubMed]
16. Susman, M.D.; Feldman, Y.; Vaskevich, A.; Rubinstein, I. Chemical deposition and stabilization of plasmonic copper nanoparticle films on transparent substrates. *Chem. Mater.* **2012**, *24*, 2501–2508. [CrossRef]
17. Li, W.; Ren, K.; Zhou, J. Aluminum-based localized surface plasmon resonance for biosensing. *Trends Anal. Chem.* **2016**, *80*, 486–494. [CrossRef]
18. Huang, X.; Tang, S.; Mu, X.; Dai, Y.; Chen, G.; Zhou, Z.; Ruan, F.; Yang, Z.; Zheng, N. Freestanding palladium nanosheets with plasmonic and catalytic properties. *Nat. Nanotech.* **2011**, *6*, 28–32. [CrossRef]
19. Reddy, H.; Guler, U.; Kudyshev, Z.; Kildishev, A.V.; Shalaev, V.M.; Boltasseva, A. Temperature-dependent optical properties of plasmonic titanium nitride thin films. *ACS Photonics* **2017**, *4*, 1413–1420. [CrossRef]
20. Gopalan, K.K.; Paulillo, B.; Mackenzie, D.M.; Rodrigo, D.; Bareza, N.; Whelan, P.R.; Shivayogimath, A.; Pruneri, V. Scalable and tunable periodic graphene nanohole arrays for mid-infrared plasmonics. *Nano Lett.* **2018**, *18*, 5913–5918. [CrossRef]
21. Agrawal, A.; Cho, S.H.; Zandi, O.; Ghosh, S.; Johns, R.W.; Milliron, D.J. Localized surface plasmon resonance in semiconductor nanocrystals. *Chem. Rev.* **2018**, *118*, 3121–3207. [CrossRef] [PubMed]
22. Wu, Y.; Niu, J.; Danesh, M.; Liu, J.; Chen, Y.; Ke, L.; Qiu, C.; Yang, H. Localized surface plasmon resonance in graphene nanomesh with Au nanostructures. *Appl. Phys. Lett.* **2016**, *109*, 041106. [CrossRef]
23. Dong, P.; Lin, Y.; Deng, J.; Di, J. Ultrathin gold-shell coated silver nanoparticles onto a glass platform for improvement of plasmonic sensors. *ACS Appl. Mater. Interfaces* **2013**, *5*, 2392–2399. [CrossRef] [PubMed]
24. Abargues, R.; Nickel, U.; Rodriguez-Canto, P. Charge dissipation in e-beam lithography with Novolak-based conducting polymer films. *Nanotechnology* **2008**, *19*, 125302. [CrossRef]
25. Shir, D.; Ballard, Z.S.; Ozcan, A. Flexible plasmonic sensors. *IEEE J. Sel. Top. Quantum Electron.* **2015**, *22*, 12–20. [CrossRef]
26. Sugavaneshwar, R.P.; Ishii, S.; Dao, T.D.; Ohi, A.; Nabatame, T.; Nagao, T. Fabrication of highly metallic TiN films by pulsed laser deposition method for plasmonic applications. *ACS Photonics* **2017**, *5*, 814–819. [CrossRef]
27. Lin, E.H.; Tsai, W.S.; Lee, K.L.; Lee, M.C.M.; Wei, P.K. Enhancing angular sensitivity of plasmonic nanostructures using mode transition in hexagonal gold nanohole arrays. *Sens. Actuators B Chem.* **2017**, *241*, 800–805. [CrossRef]
28. Chuo, Y.; Hohertz, D.; Landrock, C.; Omrane, B.; Kavanagh, K.L.; Kaminska, B. Large-area low-cost flexible plastic nanohole arrays for integrated bio-chemical sensing. *IEEE Sens. J.* **2013**, *13*, 3982–3990. [CrossRef]
29. Aouani, H.; Wenger, J.; Gérard, D.; Rigneault, H.; Devaux, E.; Ebbesen, T.W.; Mahdavi, F.; Xu, T.; Blair, S. Crucial role of the adhesion layer on the plasmonic fluorescence enhancement. *ACS Nano* **2009**, *3*, 2043–2048. [CrossRef]
30. Habteyes, T.G.; Dhuey, S.; Wood, E.; Gargas, D.; Cabrini, S.; Schuck, P.J.; Alivisatos, A.P.; Leone, S.R. Metallic adhesion layer induced plasmon damping and molecular linker as a nondamping alternative. *ACS Nano* **2012**, *6*, 5702–5709. [CrossRef]
31. Zhu, S.; Du, C.; Fu, Y.; Deng, Q.; Shi, L. Influence of Cr adhesion layer on detection of amyloid-derived diffusible ligands based on localized surface plasmon resonance. *Plasmonics* **2009**, *4*, 135–140. [CrossRef]
32. Najiminaini, M.; Vasefi, F.; Kaminska, B.; Carson, J. Optical resonance transmission properties of nano-hole arrays in a gold film: Effect of adhesion layer. *Opt. Express* **2011**, *19*, 26186–26197. [CrossRef] [PubMed]
33. Todeschini, M.; Bastos Da Silva Fanta, A.; Jensen, F.; Wagner, J.B.; Han, A. Influence of Ti and Cr adhesion layers on ultrathin Au films. *ACS Appl. Mater. Interfaces* **2017**, *9*, 37374–37385. [CrossRef] [PubMed]

34. Siegfried, T.; Ekinci, Y.; Martin, O.J.; Sigg, H. Engineering metal adhesion layers that do not deteriorate plasmon resonances. *ACS Nano* **2013**, *7*, 2751–2757. [CrossRef]
35. Cortie, M.B.; McDonagh, A.M. Synthesis and optical properties of hybrid and alloy plasmonic nanoparticles. *Chem. Rev.* **2011**, *111*, 3713–3735. [CrossRef]
36. Baburin, A.S.; Merzlikin, A.M.; Baryshev, A.V.; Ryzhikov, I.A.; Panfilov, Y.V.; Rodionov, I.A. Silver-based plasmonics: Golden material platform and application challenges. *Opt. Mater. Express* **2019**, *9*, 611–642. [CrossRef]
37. Horák, M.; Bukvišová, K.; Švarc, V.; Jaskowiec, J.; Křápek, V.; Šikola, T. Comparative study of plasmonic antennas fabricated by electron beam and focused ion beam lithography. *Sci. Rep.* **2018**, *8*, 9640. [CrossRef]
38. Arnob, M.M.P.; Zhao, F.; Li, J.; Shih, W.C. EBL-based fabrication and different modeling approaches for nanoporous gold nanodisks. *ACS Photonics* **2017**, *4*, 1870–1878. [CrossRef]
39. Vieu, C.; Carcenac, F.; Pepin, A.; Chen, Y.; Mejias, M.; Lebib, A.; Manin-Ferlazzo, L.; Couraud, L.; Launois, H. Electron beam lithography: Resolution limits and applications. *Appl. Surf. Sci.* **2000**, *164*, 111–117. [CrossRef]
40. Haynes, C.L.; Van Duyne, R.P. Nanosphere lithography: A versatile nanofabrication tool for studies of size-dependent nanoparticle optics. *J. Phys. Chem. B* **2001**, *105*, 5599–5611. [CrossRef]
41. Hicks, E.M.; Zhang, X.; Zou, S.; Lyandres, O.; Spears, K.G.; Schatz, G.C.; Van Duyne, R.P. Plasmonic properties of film over nanowell surfaces fabricated by nanosphere lithography. *J. Phys. Chem. B* **2005**, *109*, 22351–22358. [CrossRef] [PubMed]
42. Halpern, A.R.; Corn, R.M. Lithographically patterned electrodeposition of gold, silver, and nickel nanoring arrays with widely tunable near-infrared plasmonic resonances. *ACS Nano* **2013**, *7*, 1755–1762. [CrossRef]
43. Prikulis, J.; Hanarp, P.; Olofsson, L.; Sutherland, D.; Kaell, M. Optical spectroscopy of nanometric holes in thin gold films. *Nano Lett.* **2004**, *4*, 1003–1007. [CrossRef]
44. Im, H.; Lee, S.H.; Wittenberg, N.J.; Johnson, T.W.; Lindquist, N.C.; Nagpal, P.; Norris, D.J.; Oh, S.H. Template-stripped smooth Ag nanohole arrays with silica shells for surface plasmon resonance biosensing. *ACS Nano* **2011**, *5*, 6244–6253. [CrossRef] [PubMed]
45. Es-Souni, M.; Habouti, S. Ordered nanomaterial thin films via supported anodized alumina templates. *Front. Mater.* **2014**, *1*, 19. [CrossRef]
46. Yeom, S.H.; Kim, O.G.; Kang, B.H.; Kim, K.J.; Yuan, H.; Kwon, D.H.; Kim, H.R.; Kang, S.W. Highly sensitive nano-porous lattice biosensor based on localized surface plasmon resonance and interference. *Opt. Express* **2011**, *19*, 22882–22891. [CrossRef]
47. McPhillips, J.; Murphy, A.; Jonsson, M.P.; Hendren, W.R.; Atkinson, R.; Höök, F.; Zayats, A.V.; Pollard, R.J. High-performance biosensing using arrays of plasmonic nanotubes. *ACS Nano* **2010**, *4*, 2210–2216. [CrossRef]
48. Kabashin, A.; Evans, P.; Pastkovsky, S.; Hendren, W.; Wurtz, G.; Atkinson, R.; Pollard, R.; Podolskiy, V.; Zayats, A. Plasmonic nanorod metamaterials for biosensing. *Nat. Mater.* **2009**, *8*, 867. [CrossRef]
49. Hahn, C.; Hajebifard, A.; Berini, P. Helium focused ion beam direct milling of plasmonic heptamer-arranged nanohole arrays. *Nanophotonics* **2019**. [CrossRef]
50. Bhattarai, J.K.; Sharma, A.; Fujikawa, K.; Demchenko, A.V.; Stine, K.J. Electrochemical synthesis of nanostructured gold film for the study of carbohydrate-lectin interactions using localized surface plasmon resonance spectroscopy. *Carbohydr. Res.* **2015**, *405*, 55–65. [CrossRef]
51. Stiles, P.L.; Dieringer, J.A.; Shah, N.C.; Van Duyne, R.P. Surface-enhanced Raman spectroscopy. *Annu. Rev. Anal. Chem.* **2008**, *1*, 601–626. [CrossRef] [PubMed]
52. Canalejas-Tejero, V.; Herranz, S.; Bellingham, A.; Moreno-Bondi, M.C.; Barrios, C.A. Passivated aluminum nanohole arrays for label-free biosensing applications. *ACS Appl. Mater. Interfaces* **2014**, *6*, 1005–1010. [CrossRef] [PubMed]
53. Lee, S.H.; Johnson, T.W.; Lindquist, N.C.; Im, H.; Norris, D.J.; Oh, S.H. Linewidth-optimized extraordinary optical transmission in water with template-stripped metallic nanohole arrays. *Adv. Funct. Mater.* **2012**, *22*, 4439–4446. [CrossRef]
54. Zhang, X.; Li, Z.; Ye, S.; Wu, S.; Zhang, J.; Cui, L.; Li, A.; Wang, T.; Li, S.; Yang, B. Elevated Ag nanohole arrays for high performance plasmonic sensors based on extraordinary optical transmission. *J. Mater. Chem.* **2012**, *22*, 8903–8910. [CrossRef]
55. Brolo, A.G.; Gordon, R.; Leathem, B.; Kavanagh, K.L. Surface plasmon sensor based on the enhanced light transmission through arrays of nanoholes in gold films. *Langmuir* **2004**, *20*, 4813–4815. [CrossRef]

56. Lee, S.H.; Bantz, K.C.; Lindquist, N.C.; Oh, S.H.; Haynes, C.L. Self-assembled plasmonic nanohole arrays. *Langmuir* **2009**, *25*, 13685–13693. [CrossRef]
57. Rindzevicius, T.; Alaverdyan, Y.; Dahlin, A.; Hoeoek, F.; Sutherland, D.S.; Kaell, M. Plasmonic sensing characteristics of single nanometric holes. *Nano Lett.* **2005**, *5*, 2335–2339. [CrossRef]
58. Chen, J.; Shi, J.; Decanini, D.; Cambril, E.; Chen, Y.; Haghiri-Gosnet, A.M. Gold nanohole arrays for biochemical sensing fabricated by soft UV nanoimprint lithography. *Microelectron. Eng.* **2009**, *86*, 632–635. [CrossRef]
59. Live, L.S.; Bolduc, O.R.; Masson, J.F. Propagating surface plasmon resonance on microhole arrays. *Anal. Chem.* **2010**, *82*, 3780–3787. [CrossRef]
60. Ai, B.; Yu, Y.; Möhwald, H.; Zhang, G. Novel 3D Au nanohole arrays with outstanding optical properties. *Nanotechnology* **2012**, *24*, 035303. [CrossRef]
61. Luo, X.; Xing, Y.; Galvan, D.D.; Zheng, E.; Wu, P.; Cai, C.; Yu, Q. A plasmonic gold nanohole array for surface-enhanced Raman scattering detection of DNA methylation. *ACS Sens.* **2019**. [CrossRef] [PubMed]
62. Xiang, G.; Zhang, N.; Zhou, X. Localized surface plasmon resonance biosensing with large area of gold nanoholes fabricated by nanosphere lithography. *Nanoscale Res. Lett.* **2010**, *5*, 818–822. [CrossRef] [PubMed]
63. Saito, M.; Kitamura, A.; Murahashi, M.; Yamanaka, K.; Hoa, L.Q.; Yamaguchi, Y.; Tamiya, E. Novel gold-capped nanopillars imprinted on a polymer film for highly sensitive plasmonic biosensing. *Anal. Chem.* **2012**, *84*, 5494–5500. [CrossRef] [PubMed]
64. Zheng, J.; Yang, W.; Wang, J.; Zhu, J.; Qian, L.; Yang, Z. An ultranarrow SPR linewidth in the UV region for plasmonic sensing. *Nanoscale* **2019**, *11*, 4061–4066. [CrossRef]
65. Norek, M.; Włodarski, M.; Matysik, P. UV plasmonic-based sensing properties of aluminum nanoconcave arrays. *Curr. Appl. Phys.* **2014**, *14*, 1514–1520. [CrossRef]
66. Sharma, B.; Cardinal, M.F.; Ross, M.B.; Zrimsek, A.B.; Bykov, S.V.; Punihaole, D.; Asher, S.A.; Schatz, G.C.; Van Duyne, R.P. Aluminum film-over-nanosphere substrates for deep-UV surface-enhanced resonance Raman spectroscopy. *Nano Lett.* **2016**, *16*, 7968–7973. [CrossRef]
67. Kim, D.K.; Yoo, S.M.; Park, T.J.; Yoshikawa, H.; Tamiya, E.; Park, J.Y.; Lee, S.Y. Plasmonic properties of the multispot copper-capped nanoparticle array chip and its application to optical biosensors for pathogen detection of multiplex DNAs. *Anal. Chem.* **2011**, *83*, 6215–6222. [CrossRef]
68. Liu, B.; Chen, S.; Zhang, J.; Yao, X.; Zhong, J.; Lin, H.; Huang, T.; Yang, Z.; Zhu, J.; Liu, S. A plasmonic sensor array with ultrahigh figures of merit and resonance linewidths down to 3 nm. *Adv. Mater.* **2018**, *30*, 1706031. [CrossRef]
69. Lang, X.; Qian, L.; Guan, P.; Zi, J.; Chen, M. Localized surface plasmon resonance of nanoporous gold. *Appl. Phys. Lett.* **2011**, *98*, 093701. [CrossRef]
70. Ebbesen, T.W.; Lezec, H.J.; Ghaemi, H.; Thio, T.; Wolff, P.A. Extraordinary optical transmission through sub-wavelength hole arrays. *Nature* **1998**, *391*, 667. [CrossRef]
71. Park, T.H.; Mirin, N.; Lassiter, J.B.; Nehl, C.L.; Halas, N.J.; Nordlander, P. Optical properties of a nanosized hole in a thin metallic film. *ACS Nano* **2008**, *2*, 25–32. [CrossRef]
72. Jonsson, M.P.; Dahlin, A.B.; Feuz, L.; Petronis, S.; Hoeoek, F. Locally functionalized short-range ordered nanoplasmonic pores for bioanalytical sensing. *Anal. Chem.* **2010**, *82*, 2087–2094. [CrossRef]
73. Nagpal, P.; Lindquist, N.C.; Oh, S.H.; Norris, D.J. Ultrasmooth patterned metals for plasmonics and metamaterials. *Science* **2009**, *325*, 594–597. [CrossRef] [PubMed]
74. Cushing, S.K.; Hornak, L.A.; Lankford, J.; Liu, Y.; Wu, N. Origin of localized surface plasmon resonances in thin silver film over nanosphere patterns. *Appl. Phys. A* **2011**, *103*, 955–958. [CrossRef]
75. Zhang, X.; Young, M.A.; Lyandres, O.; Van Duyne, R.P. Rapid detection of an anthrax biomarker by surface-enhanced Raman spectroscopy. *J. Am. Chem. Soc.* **2005**, *127*, 4484–4489. [CrossRef] [PubMed]
76. Zhang, X.; Zhao, J.; Whitney, A.V.; Elam, J.W.; Van Duyne, R.P. Ultrastable substrates for surface-enhanced Raman spectroscopy: Al_2O_3 overlayers fabricated by atomic layer deposition yield improved anthrax biomarker detection. *J. Am. Chem. Soc.* **2006**, *128*, 10304–10309. [CrossRef] [PubMed]
77. Masson, J.F.; Gibson, K.F.; Provencher-Girard, A. Surface-enhanced Raman spectroscopy amplification with film over etched nanospheres. *J. Phys. Chem. C* **2010**, *114*, 22406–22412. [CrossRef]
78. Lee, J.; Zhang, Q.; Park, S.; Choe, A.; Fan, Z.; Ko, H. Particle-film plasmons on periodic silver film over nanosphere (AgFON): A hybrid plasmonic nanoarchitecture for surface-enhanced Raman spectroscopy. *ACS Appl. Mater. Interfaces* **2016**, *8*, 634–642. [CrossRef]

79. Bhattarai, J.K.; Neupane, D.; Nepal, B.; Mikhaylov, V.; Demchenko, A.V.; Stine, K.J. Preparation, modification, characterization, and biosensing application of nanoporous gold using electrochemical techniques. *Nanomaterials* **2018**, *8*, 171. [CrossRef]
80. Elbourne, A.; Coyle, V.E.; Truong, V.K.; Sabri, Y.M.; Kandjani, A.E.; Bhargava, S.K.; Ivanova, E.P.; Crawford, R.J. Multi-directional electrodeposited gold nanospikes for antibacterial surface applications. *Nanoscale Adv.* **2019**, *1*, 203–212. [CrossRef]

© 2020 by the authors. Licensee MDPI, Basel, Switzerland. This article is an open access article distributed under the terms and conditions of the Creative Commons Attribution (CC BY) license (http://creativecommons.org/licenses/by/4.0/).

MDPI
St. Alban-Anlage 66
4052 Basel
Switzerland
Tel. +41 61 683 77 34
Fax +41 61 302 89 18
www.mdpi.com

Processes Editorial Office
E-mail: processes@mdpi.com
www.mdpi.com/journal/processes

www.ingramcontent.com/pod-product-compliance
Lightning Source LLC
LaVergne TN
LVHW070556100526
838202LV00012B/480